STARTING DESIGN TECHNOLOGY

MICROELECTRONICS

JOHN LYNCH

Series editor: John Cave

CASSELL

Cassell Publishers Limited
Artillery House
Artillery Row
London SW1P 1RT

© Cassell 1990

First published 1990

British Library Cataloguing in Publication Data

Lynch, John, *1949–*
  Microelectronics.
  1. Microelectronics
  I. Title   II. Series
  621.381
  ISBN 0-304-31650-4

Typeset by Fakenham Photosetting Limited, Fakenham,
Norfolk

Printed and bound in Great Britain by The Bath Press, Avon

Contents

Introduction

This book invites you into the fascinating world of microelectronics. Many people think this subject is hard and a bit mysterious, but if you read on you will find yourself doing microelectronics – and enjoying it.

To get the most from this book, try not to skip pages, and have a go at all the activities. All the words that will be new to you are printed in **bold** to start with and are listed in a mini-dictionary at the back. This will give you more information and help you to answer any questions.

A new kind of revolution is taking place in our lives and it is just as important as the Industrial Revolution which began in the late eighteenth century. Maybe even more important. The 'microchip' revolution is well and truly with us.

Very complicated electronic circuits are reduced in size microscopically and built into a tiny 'chip' of silicon. Minute gold threads connect the circuits from inside their cases (known as packages) to larger pins on the outside. These we can see and use to make connections to various other components.

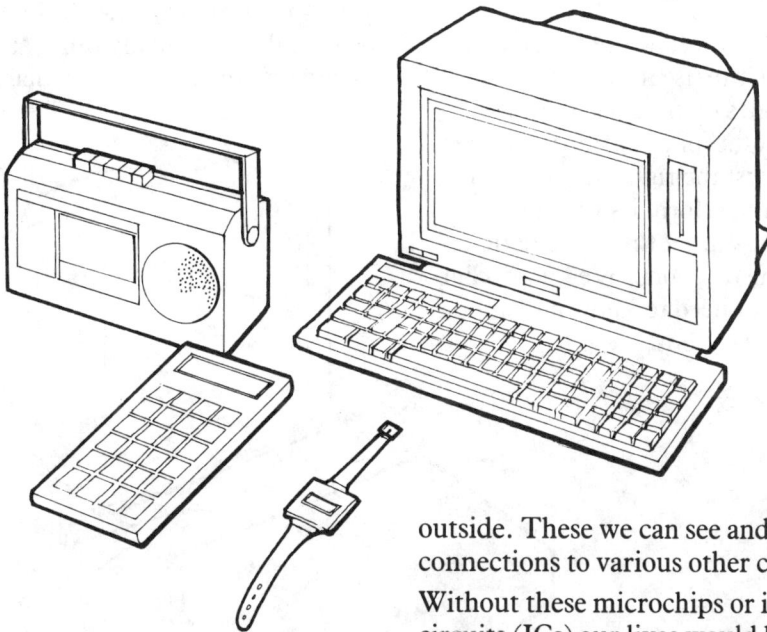

Without these microchips or integrated circuits (ICs) our lives would be very different. Think of the number of appliances, machines, gadgets, etc., that use ICs: watches, calculators, radios, computers. I'm sure you can think of many, many more.

Fact File

Some computers can carry out several hundred thousand calculations *every* second.

The Black Box

Although complicated microscopic circuits are built into tiny ICs, it does not mean that we cannot make use of them because we have only a limited knowledge of electronics. We simply treat them as **black boxes**.

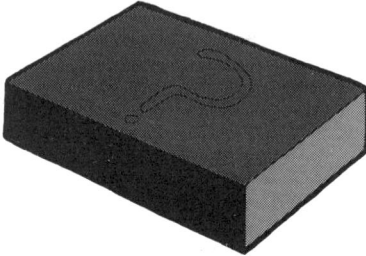

We do not need to know how the inside of a video works in order to use it. We find out what the various ones on the market can do, select the one we want (or can afford), connect it up correctly, adjust the right buttons and knobs and it should perform its various functions properly. We get this information from the book of instructions. We know what it does but we do not need to know how it does it. We treat it as a black box.

The same applies to ICs. We find out what they are capable of doing, choose the ones we want, connect them up correctly, make the fine adjustments and they should also perform their functions properly. This information or data we gather from instruction booklets or even the catalogues from which we buy the ICs. So anyone with an interest, the data, and patience can use microchips to build some very exciting and useful projects.

You will find it useful to read or refer to the **Electronics** book from this series, as many methods and components used in this book apply to electronics as well. References are made to the book many times as a way of learning more about a particular topic.

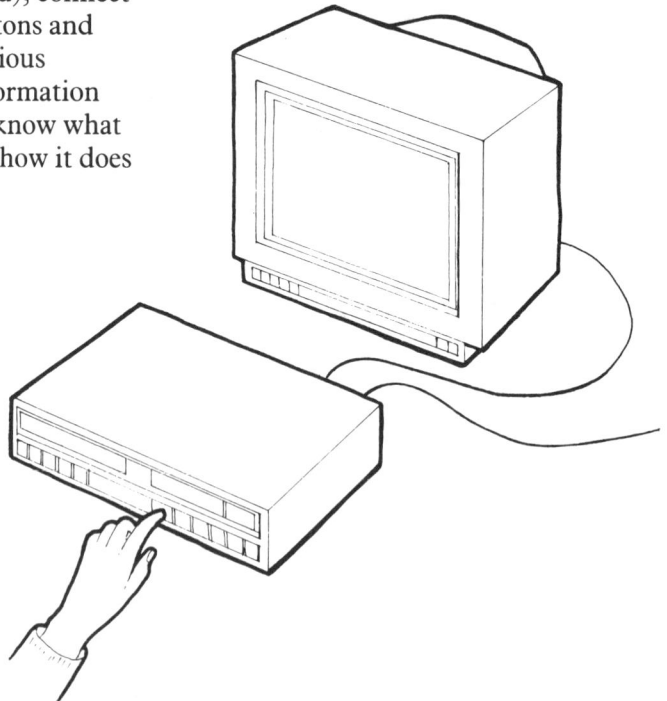

1 Packages

The package or container that an IC comes in can vary enormously in looks but the **dual in line** (DIL) package is probably the most common and easily recognized by almost everyone.

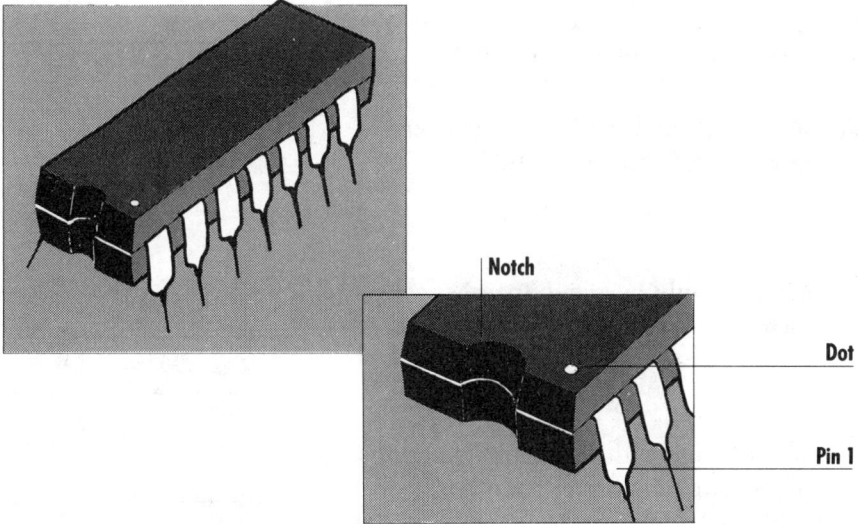

It varies in size and number of pins but certain similarities can be seen even between those produced by different manufacturers.

The pins are normally 2 mm apart and the rows of pins 6, 8, 10 mm, etc., in length. There should be a notch or small dot on the top surface identifying where pin 1 lies.

The pins are numbered in anticlockwise order from pin 1 to however many pins the package has.

Packages

Connections

Armed with these basic rules we can come closer to understanding the pin connections on a diagram. The use of diagrams makes it easier to see the vital information that we need for connecting up. Most diagrams are drawn as though looking from above. A pin layout for a typical 14-pin DIL IC would look like this.

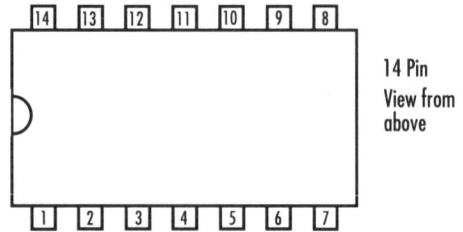

14 Pin
View from
above

Connections to and from the IC and between other components are shown as solid lines.

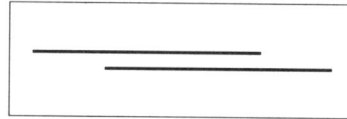

Where the lines should be joined together a blob represents the joint.

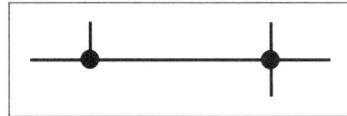

Where two lines cross without a blob it simply means that one wire passes over the other without touching.

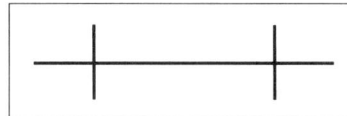

The solid line at the top of a diagram with *Vcc* or +Volts printed on it is the track or wire going to the positive (+) of the power supply.

The solid line at the bottom of the diagram labelled with 0 volts, earth, gnd or ⏚ is connected to the negative (−) of the power supply.

2 Basics

Integrated circuits, although capable of performing some marvellous feats, are virtually useless on their own.

For instance, a calculator IC can work out very difficult problems in a fraction of a second but on its own you would never know it was doing it.

The actual output is only very small electrical pulses which of course we can't see, hear or touch. They need to be made bigger or changed into something else.

Other components are needed in order to put these tiny outputs to good use.

Before we look at some of the different components let's briefly go through what electronics is about.

Millions of invisible particles called electrons are pushed round a circuit by the pressure from a power supply. This supply is very often a battery of some kind. The amount of pressure is given in volts.

A single cell has 1.5 volts.

A battery is made up of several cells connected in series although they are hidden in the shell or case. Common values are 4.5 volts, 6 volts, 9 volts and 12 volts.

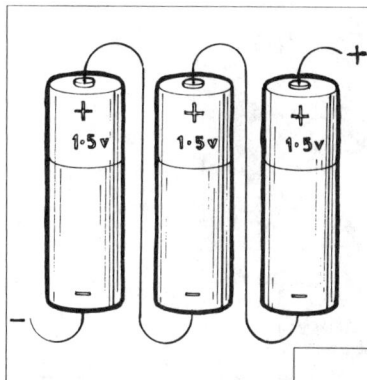

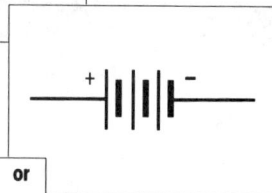

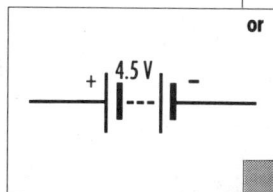

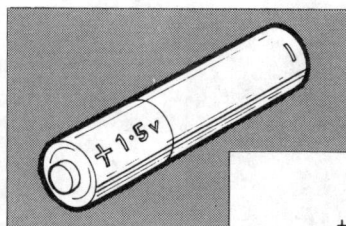

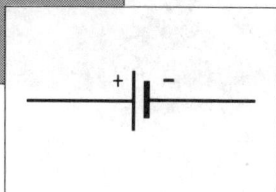

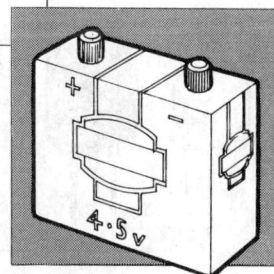

A cell or battery has two terminals, positive (+) and negative (−). Electrons flow from the negative through the circuit to the positive terminal.

It used to be believed that they flowed in the opposite direction and we still mostly use that theory today. We call this conventional current.

Components

Here are some components that we need to know about, together with their symbols.

Some merely alter the flow of electrons in some way.

Resistors slow down the flow.

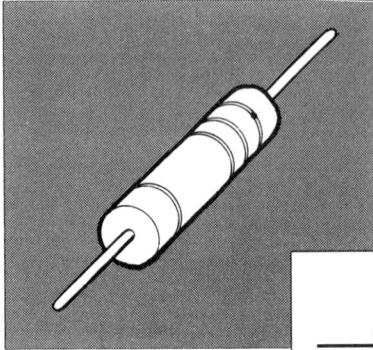

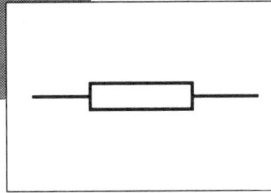

Capacitors store and release it.

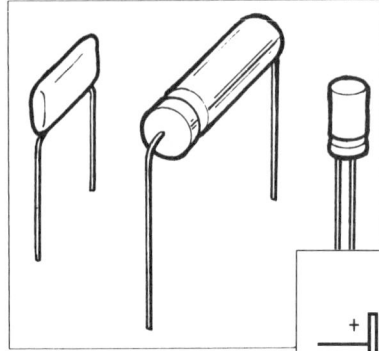

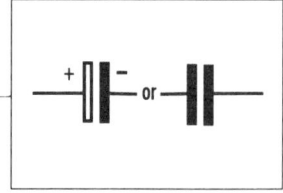

Transistors control it.

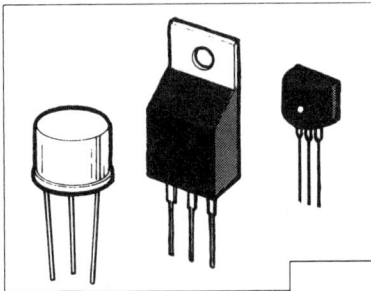

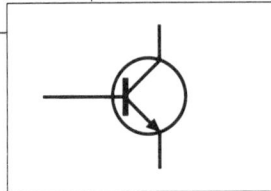

Diodes are one-way streets for electrons.

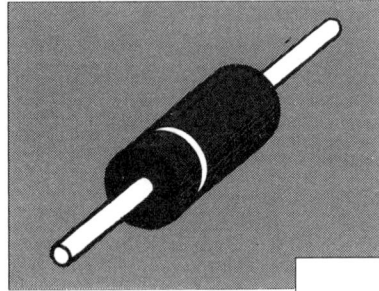

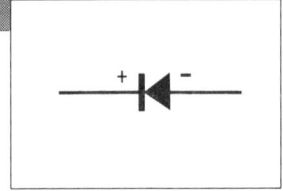

Components

Others can change electrical energy into a
different form of energy.

Bulbs turn electrical energy into light energy.

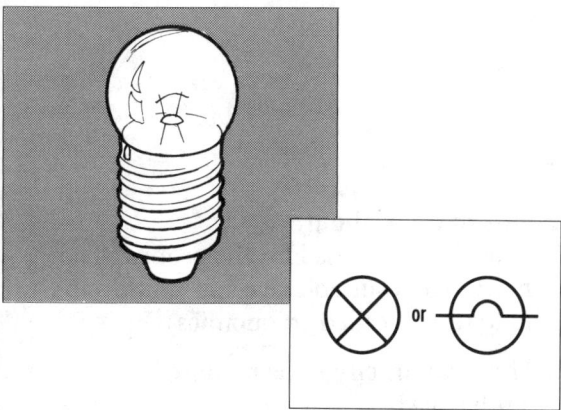

Light-emitting diodes (LEDs) light up when
current flows.

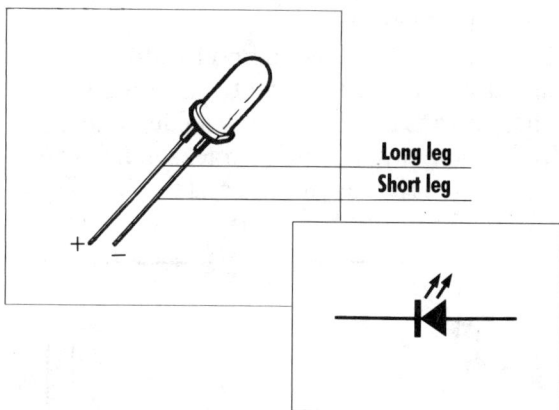

Long leg
Short leg

+   −

or

Loudspeakers change electrical energy to
sound energy.

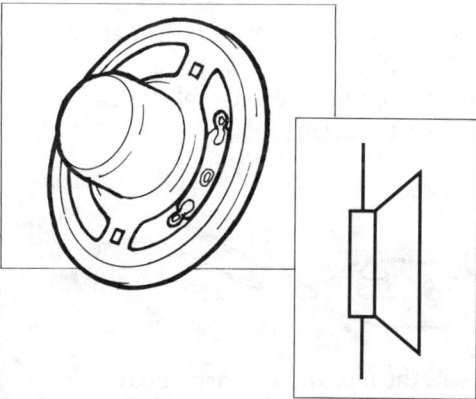

Motors change electrical energy to moving
energy.

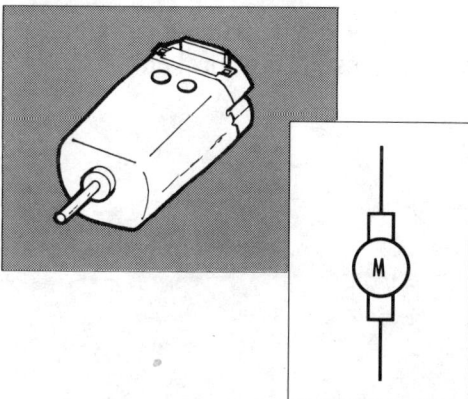

M

Other components will be explained as they
appear in different projects.

11

Making a Continuity Tester

This project serves two very useful purposes.

1 It will help test and find faults in later circuits.
2 It can be used around the house to test bulbs, fuses, etc.

The circuit is designed to find breaks in a circuit where electrons can't get through. These breaks could be poorly soldered joints (called dry joints), hairline cracks in tracks on a circuit board, blown fuses and many more.

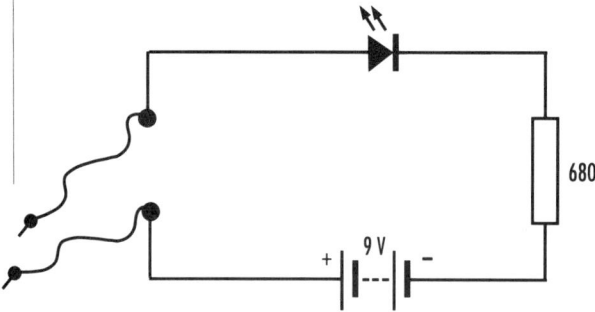

As you can see, the circuit will not work unless whatever we are testing is unbroken.

Points to think about:

1 What could be used for the probes? They need to be a suitable size and comfortable to hold (this is called ergonomics).

2 How can the circuit be mounted? In a box or a tube maybe?

3 The LED must be placed so that it can be easily seen without having to move your eyes or head too far (this is also ergonomics).

4 The easiest way to connect to the battery is to use battery snaps. These have a red lead which goes to positive and a black lead which goes to negative.

+ **Red positive**

− **Black negative**

5 How will the circuit be constructed?

6 The printed circuit board (PCB) method is most suitable for later projects. Using it now on a simple circuit will give you some practice.

3 Circuit Construction

Making a Printed Circuit Board (PCB)

This way of making circuits is the most common in electronics. Components are soldered between thin copper lines, or tracks, on a board instead of connecting wires. We shall be using **surface mounting**, which is explained later.

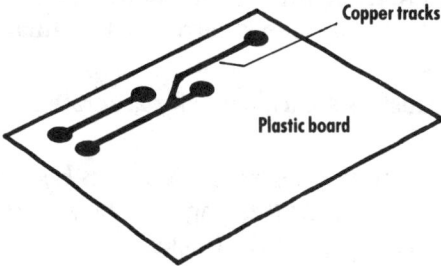

To make a PCB, you will need:
1 Copper-clad board (40 × 40 mm). This is very thin copper sheet bonded to a plastic board.
2 A good spirit-based felt-tipped pen (medium tip).
3 Ferric chloride solution. **This is harmful. It should only be used when a teacher is in charge and you are wearing goggles.**

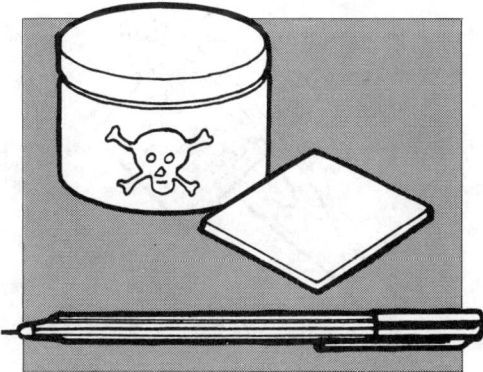

First mark out the tracks on the copper side of the board with pencil and then with the felt-tipped pen. The ink lines should be 2 to 3 mm wide and have fatter round ends. The components are soldered at these points, so more copper makes it less likely for the components to be pulled off the board.

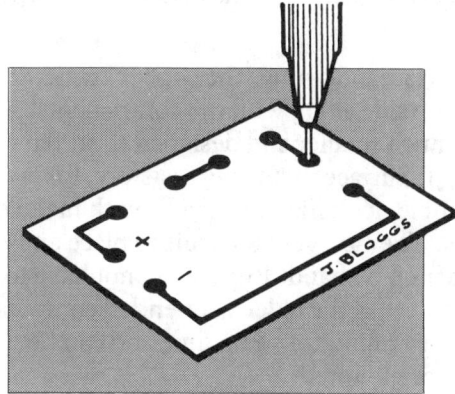

Next, the board is put into a *warm* ferric chloride solution. All the copper *not* covered by ink is eaten away. When ready, the board is washed and the ink removed with a solvent or wire wool.

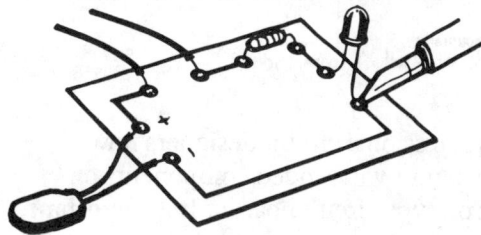

Finally, the LED and resistor legs are bent and tinned (together with the track ends) and soldered to the board.

Circuit Construction

Surface Mounting

Method of the future: Surface mounting

Up to a few years ago, most PCBs had components on top of the board with their legs going through holes and soldered to copper tracks underneath. This held them securely in place.

In surface mounting, the copper tracks are on the top side along with the components, which are much smaller and designed to sit flat on the board. Surface mounting resistors, for example, look like tiny bricks with metal end-caps. They are very difficult to place and solder on by hand. But you do not have to worry yet as the older wire-ended components will be around for some time and can be surface mounted!

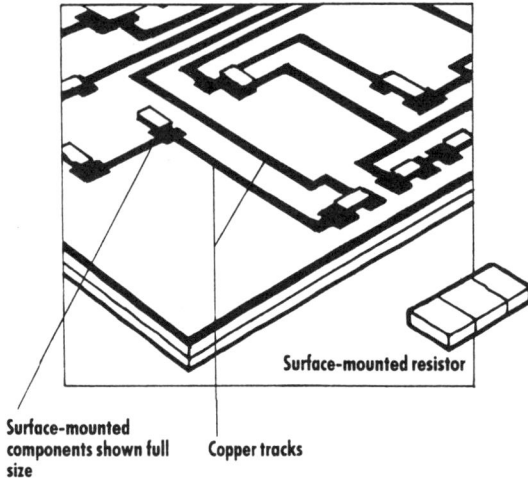

Surface-mounted resistor

Surface-mounted components shown full size   Copper tracks

Many professional circuit designers now surface mount wire-ended components on their prototype circuit boards. It is important to make sure, however, that larger components are secure.

Hints for Making Circuits

Make sure that your circuit diagram is correct. Your teacher should be able to advise.

A neat, tidy, well-spaced-out circuit should be your aim. Avoid cramming components together where possible.

If you are using the PCB method, double check that your plan is correct before final etching.

Dry transfers are available for accurate IC pads.

Use a continuity tester on your PCB *before* you start soldering. At this stage it is easy to find 'invisible' breaks in the tracks.

Once you have soldered the components, check that they are in the correct place and the right way round *before* you switch on.

Fact File

An IC containing thousands of components could fit through the eye of a needle.

Soldering

Solder is an alloy (mostly lead) that can be melted with a hot soldering iron to form a joint.

Holding the soldering iron in one hand and a length of solder in the other, heat up a track end (**pad**) and touch the solder to it. As the solder melts, flux will flow out of its centre, helping the solder to run and stick to the pad. It is useless to melt the solder on to the iron tip as the flux will disappear straight away as smoke.

This coating with solder is called tinning. You must also tin the bottom of the legs of components and stripped ends of wires. If you haven't got three hands, hold the component or the stripped end of wire with a bulldog or crocodile clip. This serves two purposes:
1. It acts as a heat sink to take away much of the heat which would otherwise run up the component leg and cause damage.
2. It prevents you from burning your fingers. Plasticine is useful for wire, or components which are not so delicate.

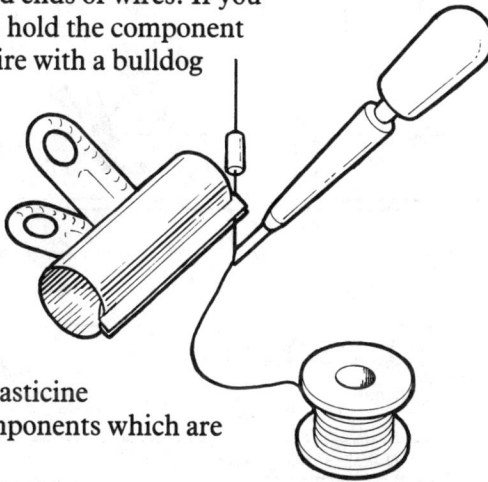

Once your circuit is complete the next thing to do is to test it.

1 Look first to see if you have soldered the components in their right positions. Check that they are the right way round (LED and battery leads).
2 Closely inspect the soldering joints.
3 If all looks well, connect the battery and touch the probes together. The LED should of course light up.

Now is the time to put the circuit in its package and start continuity testing.

4 Integrated Circuit Families

There are two main families of integrated circuit available on the market and in general use. Each has advantages and disadvantages. The two types are **Transistor Transistor Logic** (TTL) and **Complementary Metal** **Oxide Semiconductor** (CMOS, pronounced sea moss).

Although they will perform the same job in a circuit, if we consider their differences, a careful choice has to be made.

The Big Match
A five-round contest between two types of integrated circuit to find the most suitable one for general school and project work.
On the left we have CMOS. On the right we have TTL.

	CMOS		TTL
ROUND 1	Works on any voltage supply between 5 and 15 volts	SUPPLY VOLTAGE	Only works on a 5 volt supply
ROUND 2	Can switch on and off 25 – 50 million times per sec	SWITCHING SPEED	Can switch on and off 30 times faster
ROUND 3	Prone to static damage; extreme care needed	DAMAGE BY STATIC ELECTRICITY	Not prone to static damage
ROUND 4	Must be connected to + volts or 0 volts	UNUSED INPUT PINS	Can be left 'floating'
ROUND 5	Uses very little power (electricity)	POWER	Uses much more power

Judge's Decision: A close contest but, because it will work on almost any battery that can be bought in a shop and uses little power, the winner is CMOS.

Earthing

The biggest disadvantage with CMOS comes in round three. It can be damaged by static electricity and much care is needed in using it.

To help overcome the problem:
1 Only handle CMOS ICs when absolutely necessary.
2 Keep them in their anti-static cases for as long as possible.
3 Earth yourself to get rid of any static in your body by, say, touching a metal tap on the sink before you touch the IC.

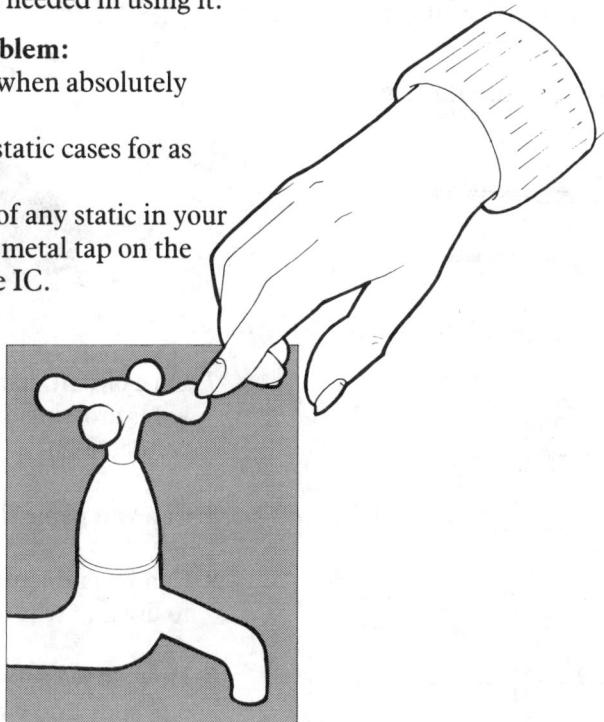

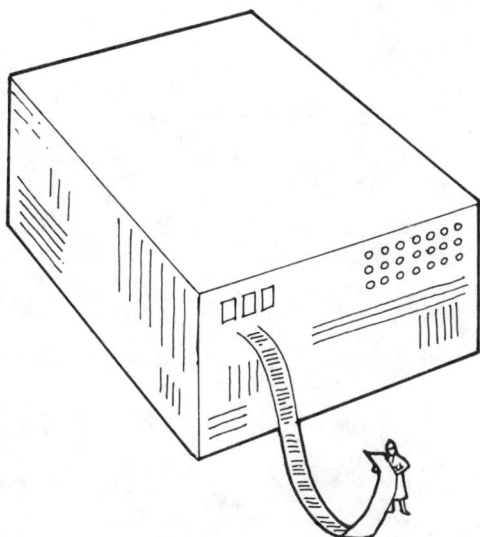

Fact File

Twenty years ago a typical computer would have filled an area the size of a classroom from floor to ceiling. It would also have needed to be kept at a constant temperature and humidity. Today a machine with the same amount of power would comfortably fit on your lap.

Questions

Test Yourself So Far

Look back over what you have read and use the mini-dictionary to help you answer these questions.

1 Name two ways to identify the legs on an LED.

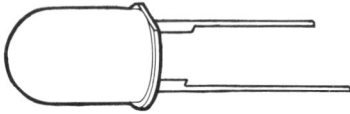

2 Will the LED in this circuit light up? Explain.

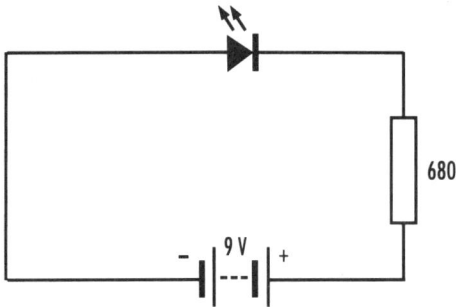

3 How do you identify the leads on battery snaps?

4 What is a black box?

5 What does 'dual in line' (DIL) refer to?

6 Why is there a notch or dot on the top of an IC?

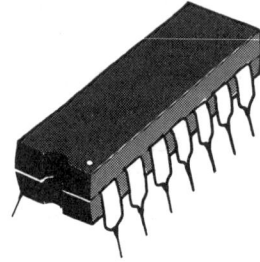

7 On a circuit diagram, how do we show two wires crossing over each other but not connected?

8 Can you name the two families of IC?

9 Can you remember which one we are going to use and why?

10 How do we get rid of static electricity from our body and why do we need to?

5 Logic ICs

A computer, calculator or digital watch works by a simple method of switching on and off. Thousands of parts of a device can switch on and off at different speeds and for different lengths of time.

Each individual switch action (on or off) is a piece of information called a **bit**. Eight pieces of information is called a **byte**. A thousand bytes is called a **kilobyte**, a million bytes is called a **megabyte**. You may have heard these terms when computer memory is being discussed.

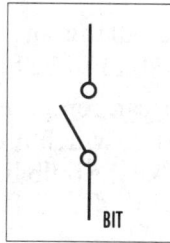

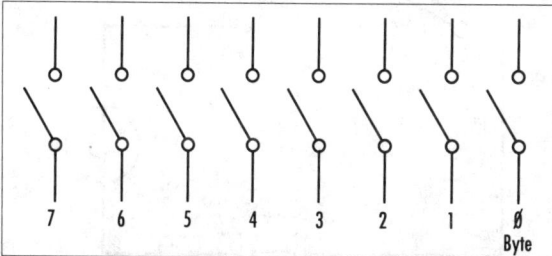

The first principle that needs to be understood is the term **logic level** or **logic state**.

Logic level refers to a point in a circuit which must be either

Positive or **Negative**

1 or **0**

High or **Low**

Nothing in Between

This is the basis of any **digital** system.

BIT

7 6 5 4 3 2 1 Ø
Byte

Logic Gates

A **logic gate** is a circuit which has inputs and
an output. Inputs can be either logic 1 or logic
0. An output can be either logic 1 or logic 0.

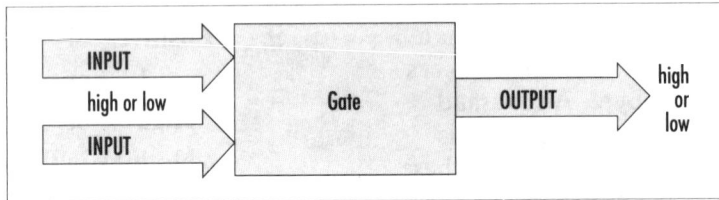

The different types of gate do different jobs on
inputs and produce different outputs.

Let's look at a simple gate called an **AND
gate.**

Here is its symbol. It has two inputs, A and B,
and one output, Q.

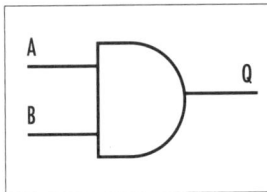

In order to make the output high we must
make *both* inputs high. If either or both inputs
are low the output will be low.

We call this an AND gate because inputs A
and B must be high to give a high output.

We can compare this with two switches in
series; switch A *and* switch B must be on to
make the bulb light up.

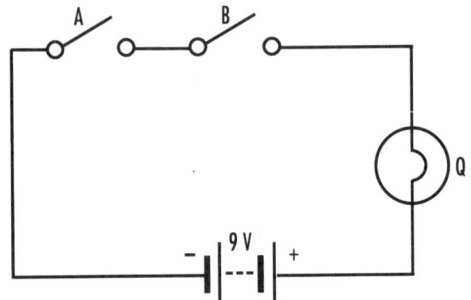

Logic Levels, Inputs, Outputs

In reality, a logic gate is a very complicated circuit within an IC and we therefore treat it as a black box.

A typical logic IC could have several gates. A 4081 has four AND gates. It is shown here with its pin connections.

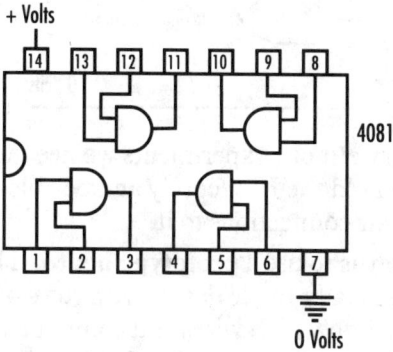

A good way to learn what the different gates can do is by experiment.

We need to know two important things:
1 How to set inputs at logic levels 1 or 0.
2 How to read the output logic level.
To set inputs we simply connect them to either the positive supply rail or the negative supply rail.

To read the output logic level we can use an LED and series resistor and a pair of leads as a very simple **logic probe**.

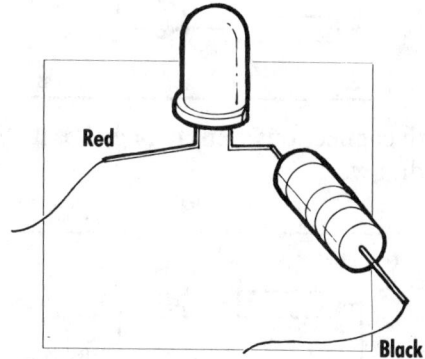

In order to make an LED light up it must be connected the right way round. The positive side must be connected to + of the supply (logic level 1) and the negative to − of the supply (logic level 0).

If we connect a red wire to the positive side and a black wire to the resistor on the negative side it will be easier to tell them apart quickly.

Conventional current flows from positive through the LED and resistor to negative.

Prototype Board

If both connections went to positive it wouldn't work.

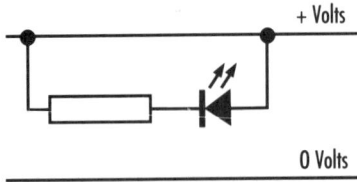
+ Volts / 0 Volts

If both connections went to negative it still wouldn't work.

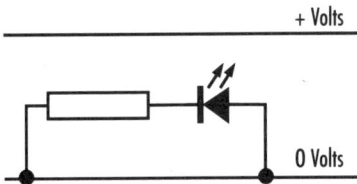
+ Volts / 0 Volts

So if we connect the black lead to 0 volts and the red lead to the output of a gate, the LED will only light up when the output is high.

When this happens we say the output is **sourcing** current. In other words it is supplying current to the LED.

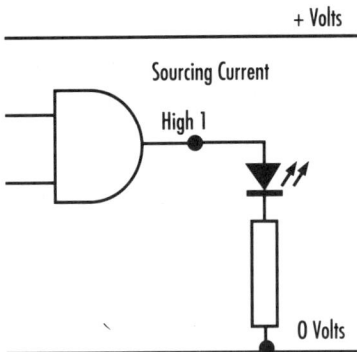
+ Volts
Sourcing Current
High 1
0 Volts

If we were to connect the red lead to + volts and the black to the output of a gate, the LED would only light up if the output was low.

When this happens we say the output is **sinking** current. Current is flowing from + volts through the LED and resistor *into* the gate.

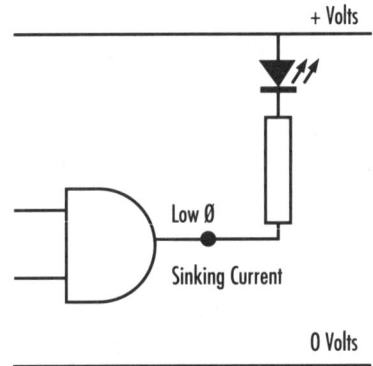
+ Volts
Low Ø
Sinking Current
0 Volts

To carry out our experiments we need to be able to hold the IC securely and be able to make our connections to it.

One way is to use a **prototype board**. This is a professionally-made device with rows of sockets into which wires and components including ICs can be inserted. The wire must be of the single core type. All the sockets in the same row are connected together underneath the board. The long rows at the top and bottom of the board can be used as the + volts and 0 volts supply rails. The vertical rows are divided in the middle by a channel to give two sets, one top and one bottom. An IC would be placed so that the channel separated the two rows of pins.

These purpose-made boards are also useful for trying out a circuit before you make a permanent one. This is called circuit modelling.

Alternatively we can make our own.

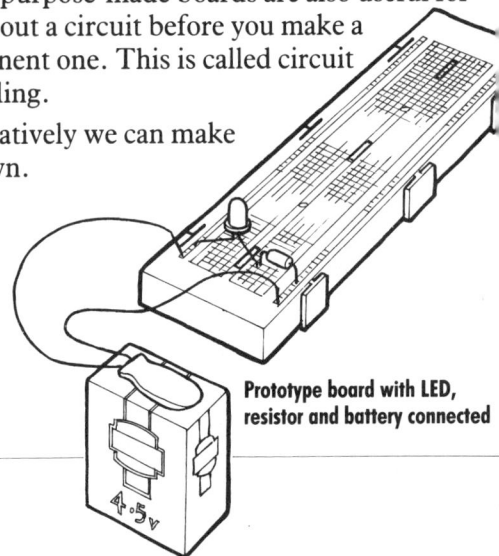

Prototype board with LED, resistor and battery connected

Making a Logic Board

Design and make a logic board which will hold a 16-pin DIL IC. (We need only 14 pins for logic work but 16 will be useful in later projects.)

Points to think about:

1 The IC will need to be easily replaced to test different types.
2 IC sockets are available. They need to be securely mounted.

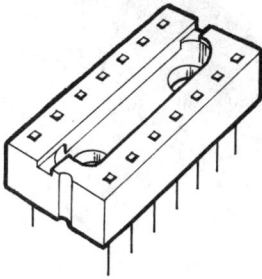

3 Connections to the IC need to be easily made and changed.
   - Small plugs and sockets?
   - Crocodile clips?
   - Twisting wire round?

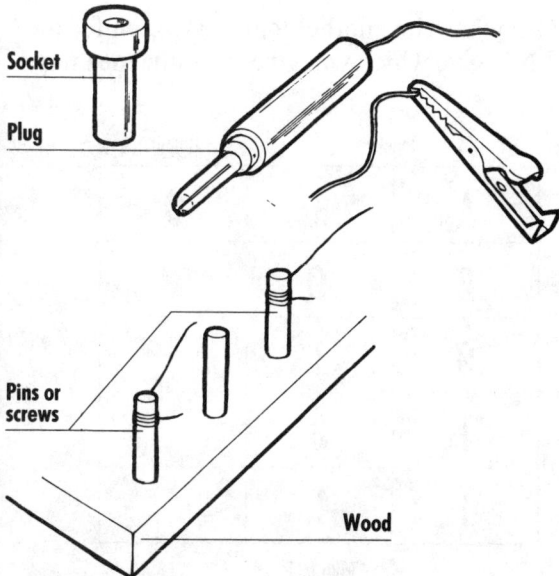

Socket

Plug

Pins or screws

Wood

4 The pins on an IC are very close together (2 mm) so it would be very easy to connect adjacent pins accidentally.
5 We need to be able to spread out or 'extend'.

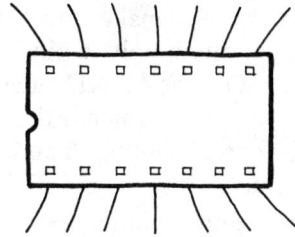

- Printed circuit board?

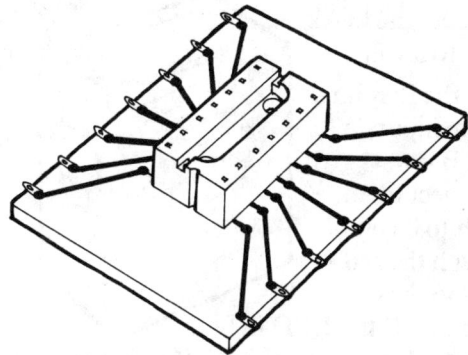

- Compression joints on wood?
- Matrix board and pins?

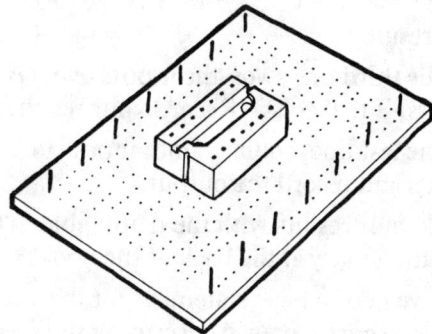

Perhaps you can think of others.

23

6 Logic Gate Experiments

The AND Gate

Experiment 1

Using a prototype board or our own logic board, insert a 4081 IC correctly. (Don't forget to earth yourself first.)

Use the IC diagram to make your connections. Connect the + volts supply to pin 14 and the 0 volts to pin 7. We have 4 AND gates to choose from. Let's use gate 1, which is the top left. Pins 13 and 12 are the inputs A and B and pin 11 is the output Q.

Record your results by filling in the output (Q) column of the *truth table*. Using an LED/resistor probe, connect the black lead to 0 volts.

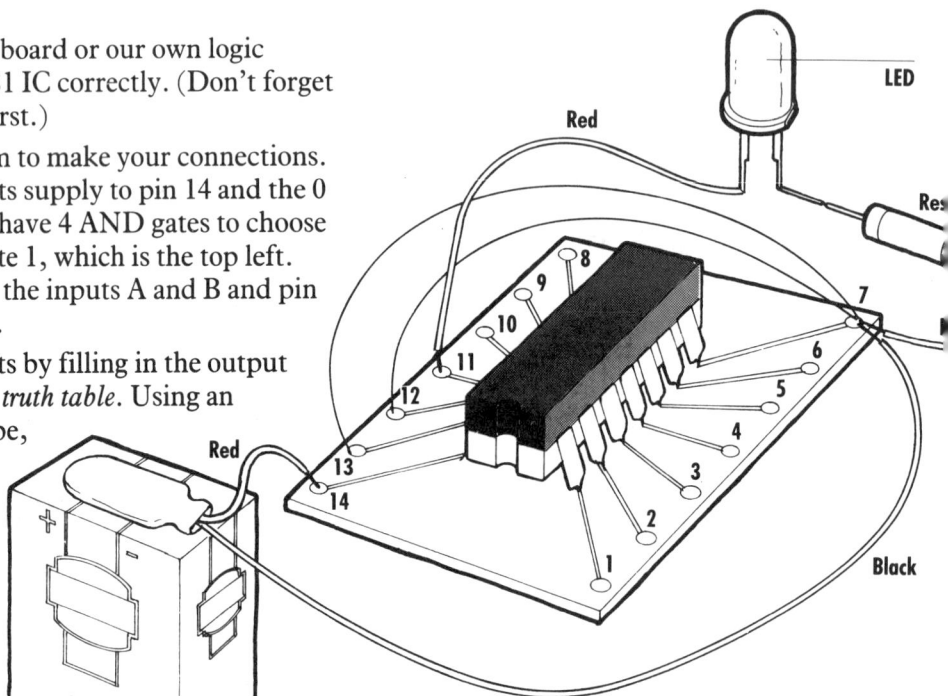

For the first line, make inputs A and B low (connect them both to 0 volts). Touch the red lead to the output (pin 11). If the LED lights, insert a 1 in the Q column. If it doesn't light, insert a 0.

For the second line, make input B high (connect it to + volts). Leave A low. Record your result.

For the third line swop the inputs over (A to + volts, B to 0 volts). Record your result.

For the last line, connect both inputs to + volts and record the output.

Check your results with the truth table in the mini-dictionary at the back of the book.

Now we need a logic statement for the gate. This is a short sentence describing the results. One statement could be, 'The output is high *only* when both inputs are high'. The logic statement should be so precise that you could fill in a truth table just by reading it.

Can you write another logic statement for the AND gate which will give the same meaning?

Inputs		Output
A	B	Q
Ø	Ø	
Ø	1	
1	Ø	
1	1	

Truth Table

The Logic Probe

You may experience a problem with using your probe. If the LED lights up you can be quite certain that the output is high. If the LED doesn't light up it could be either a low output or a bad connection when you touch the probe to the output pin.

An improved version of the logic probe would be more accurate and interesting to use and also very useful for testing later circuits.

We use two LEDs, one red and one green. If the probe detects a high output the red will light. For a low output, green will light. If there is a bad connection, both LEDs will be off or glow dimly.

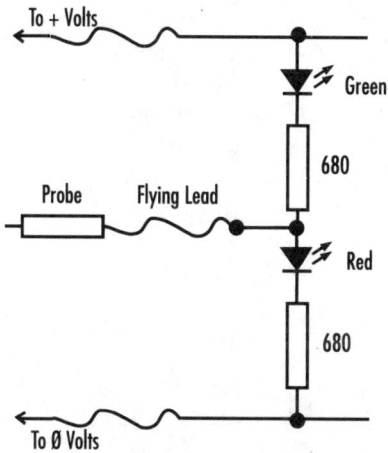

The circuit is very simple, as you can see.

If the output is low, the gate will sink current and the green LED will glow brightly. The red can't light up because it will have a 0 at both ends.

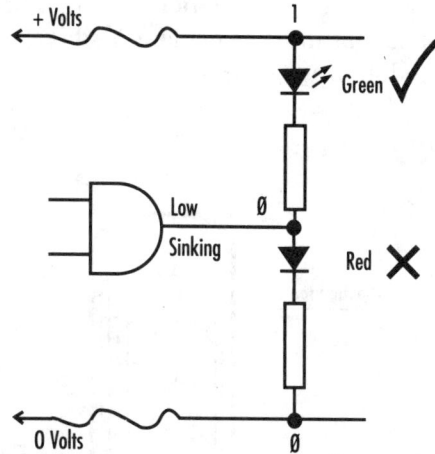

If the output is high, the gate will source current and the red LED will glow brightly. The green can't light up because it will have a + at both ends.

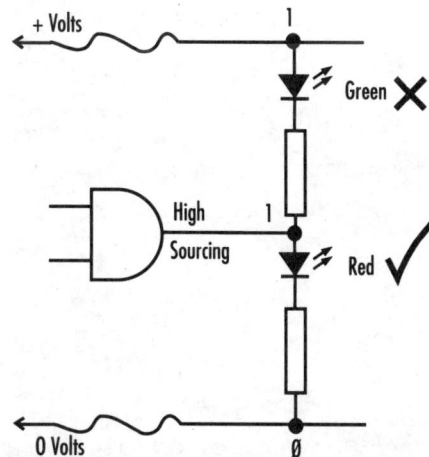

Fact File

Texas Instruments produced the first silicon integrated circuit in 1958.

Making a Logic Probe

Design and make a logic probe which will give an easy-to-read LED display of logic levels.

Points to think about:

1 The probe must share the same power supply as the circuit under test so suitable connections must be made. Possibly crocodile clips again?

2 Many of the decisions that had to be thought about for the continuity tester also apply to this project.

3 Perhaps the logic probe could be incorporated with a continuity tester.

4 The circuit could be made very small so as to fit into a thin tube. Miniature LEDs are available.

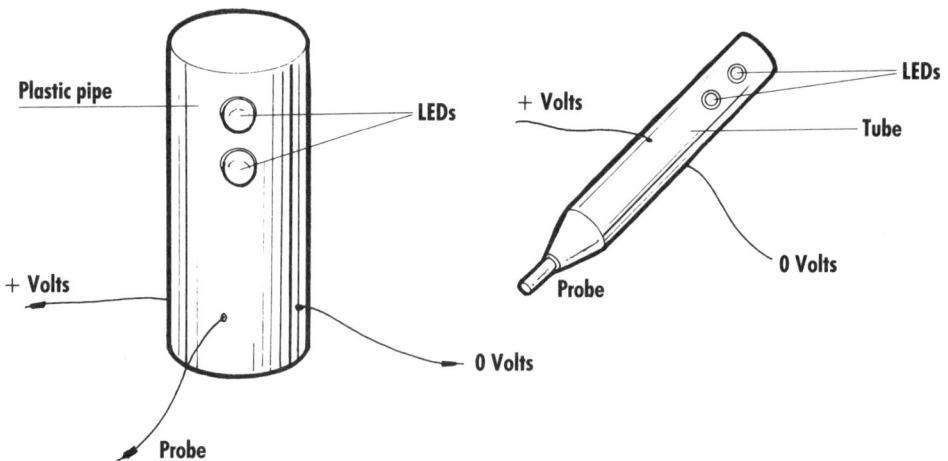

Fact File

Computers are said to be intelligent. A human has an average IQ of one hundred. The best computer would compare with a half-baked maggot.

NAND, OR and NOR Gates

Experiment 2

The next type of gate to explore is the **NAND gate**.

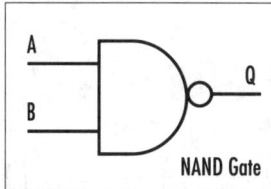

NAND Gate

Set up a 4011 IC as before and repeat your experiment.

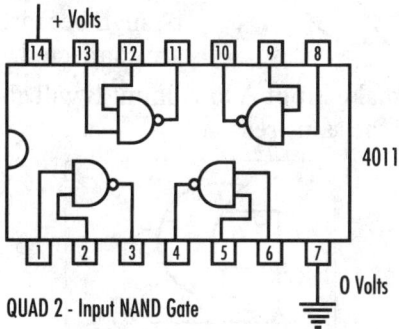

QUAD 2 - Input NAND Gate

Check your results against the truth table in the mini-dictionary.

Write a suitable logic statement for the NAND gate.

Compare the outputs with the AND gate results. What do you notice?

Experiment 3

The next gate is called an **OR gate**.

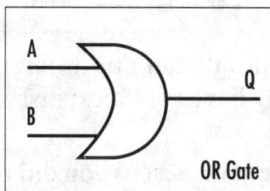

OR Gate

Set up a 4071 as before and carry out the tests.

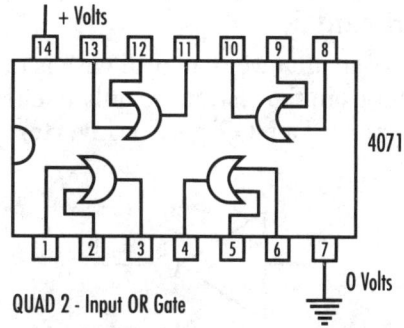

QUAD 2 - Input OR Gate

Check the results and write a suitable logic statement.

Experiment 4

The **NOR gate** is the last of the two input gates that we shall test.

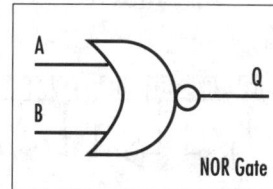

NOR Gate

Set up a 4001 as before and carry out the same procedure.

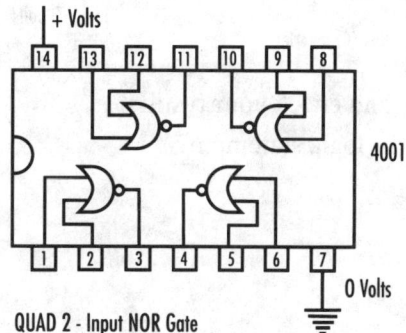

QUAD 2 - Input NOR Gate

Record and check your results and compile a logic statement.

Compare the NOR gate to the OR gate. What do you notice?

Logic Gate Experiments

The NOT Gate

Experiment 5

The last gate that we shall deal with is rather different from the rest: it has only one input (A). It is called a **NOT gate** or **inverter**.

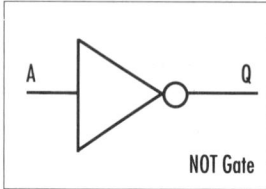

NOT Gate

As it has only one input, its truth table has just two lines.

Also, with one input the IC has room for six gates.

Set up a 4069 IC and use Gate 1 (pin 13 as the input, pin 12 as the output).

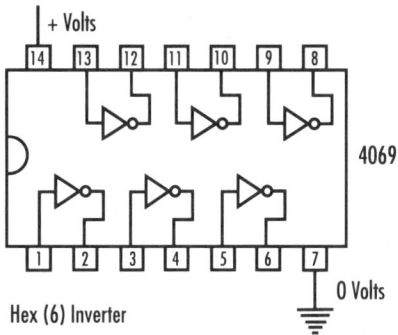

Hex (6) Inverter

Record and check your results.

Write the logic statement.

Experiment 6

Set up the NOR gate IC (4001) again.

Connect the two inputs of Gate 1 together.

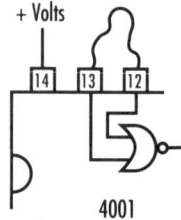

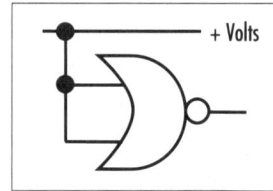

4001

If you make input A high, input B will also be high because they are connected.

If you make input A low, input B will also be low for the same reason.

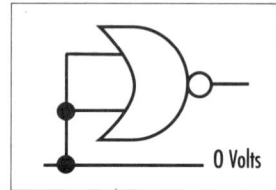

You can't make the inputs different. If you try, you will 'short' the battery because you will be connecting + volts to 0 volts. You could also damage the IC.

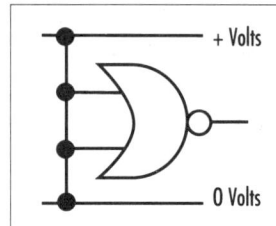

Because the inputs can't be different, the truth table can only have two lines; just like the NOT gate.

Carry out the same test as you did on the inverter.

What do you notice?

Combining Gates

For the next two experiments you should use either a prototype board or join with a friend and use both logic boards, as we are going to connect more than one IC.

Experiment 7

Set up an AND gate (4081) and NOT gate (4069 or make up your own from a NAND or NOR gate). You must use the same power supply for both ICs.

Connect the output of the AND gate to the input of the NOT gate.

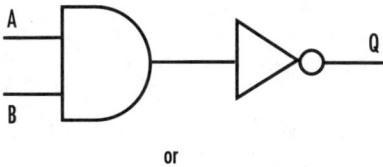

or

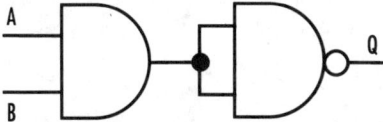

Draw the truth table with inputs to the AND gate as A and B and the output of the NOT gate as Q.

Record your results.

What do you find?

If your connections are correct you should have created your own NAND gate.

By connecting a NOT gate and an AND gate in this way we have made a NOT/AND gate which is how the NAND gets its name (N'AND).

Experiment 8

Repeat Experiment 7 using an OR gate.

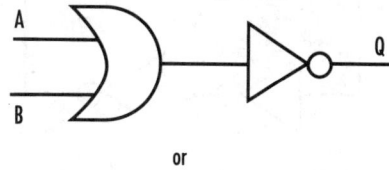

or

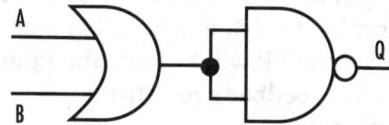

You should now discover how the NOR gate gets its name.

You should by now be aware that there are thousands of ways of joining the inputs and outputs of different logic gates to get any result that is needed. Calculations by computers and calculators are carried out in this way.

Logic Gate Experiments

Making an Electronic Switch

This is a very sensitive electronic 'latching' switch.

This circuit uses two NOT gates. The output of one goes to the input of another and a high value resistor is used as a feedback.

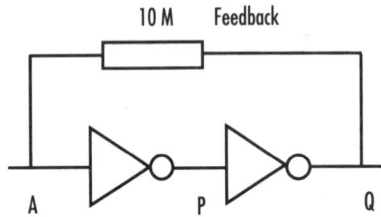

If the input of the first gate (A) is momentarily made low by 'bridging' with the finger, we know that point P will be high and point Q will be low. The **feedback resistor** holds or latches the input low.

If the input is now made high, it will latch in that position.

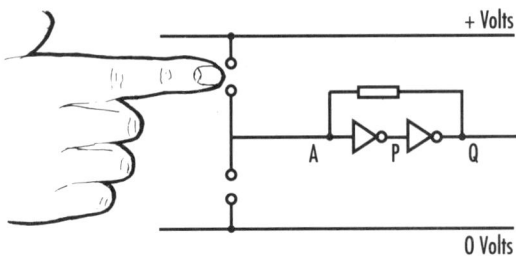

We could use different coloured LEDs at outputs P and Q. When one comes on, the other goes off. Could this circuit be adapted as a logic probe?

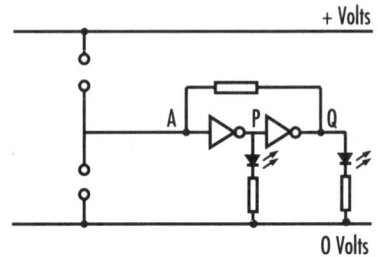

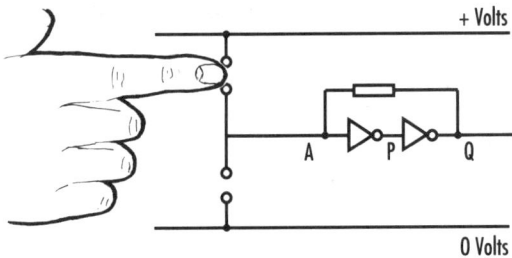

Making an Electronic Switch

If you wanted to use this switch as a door entry signal, say, for your bedroom, you could use a buzzer instead of an LED but you would need to 'amplify' the output from the IC.

For this a transistor is needed. (*See* **Electronics**.)

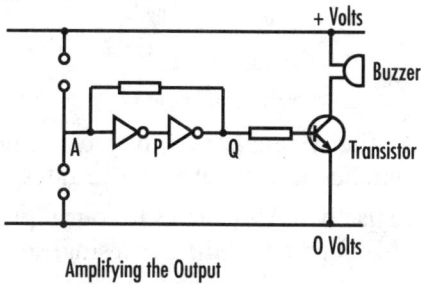

Amplifying the Output

Points to think about:

1 How will you construct the circuit – PCB, matrix, etc.?
2 Thinking back to the CMOS/TTL fight we must not let unused gate inputs 'float'. They can be connected to the nearest supply rail. If they float, unnecessary power will be wasted.
3 This must be done on all circuits using logic ICs.

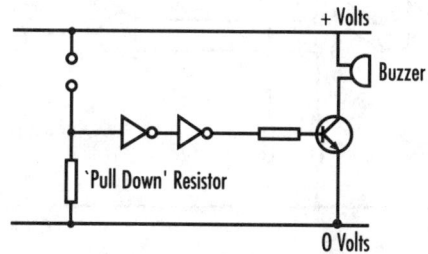

4 What could you use for bridge contacts?
   ● Metal foil? Drawing pin heads?
   ● Membrane panel? (*See* **Electronics**.)
   ● For a door switch do you really want it to latch?

If not, don't use the 10M resistor as a feedback. Connect it between the input and 0 volts to hold the input low. This is called a **pull-down resistor**. Your finger bridge will overcome the pull-down effect because your body resistance is less than 10M. When you release your finger, it will spring back to 0 volts.

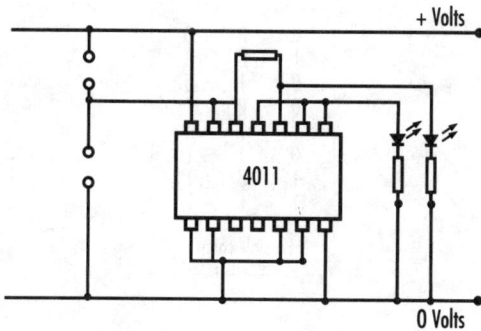

The latching circuit could be used for a variety of games where one false move latches the buzzer on; e.g. shaky-hand game, ball-bearing maze?

Logic Codes

A variety of gates could be combined to create a system needing a certain code to break it. A common use for such a device is a security lock.

In the first example, by using two switches with two pull-down resistors because an AND gate is being used, *both* switches must be in the on position in order to give a high output. The code is therefore on, on or 1, 1.

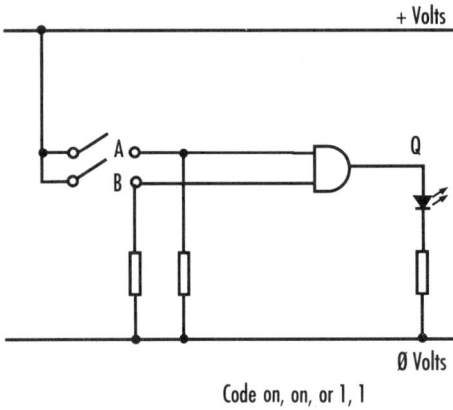

Code on, on, or 1, 1

In the second example, by changing one of the pull-down resistors into a **pull-up resistor**, switch A must be *off* and switch B must be *on* to give a high output. The code is therefore off, on or 0,1.

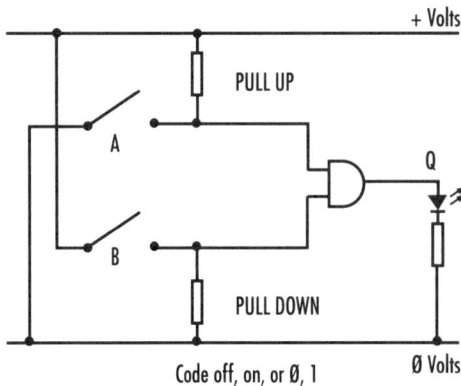

Code off, on, or 0, 1

There can be only four different codes for two input switches. Where have you seen these codes before?

OFF	OFF
OFF	ON
ON	OFF
ON	ON

4 possible codes

To make the system harder to crack we need more switches and therefore more gates.

Here we use two AND gates for our inputs and a third AND gate to reduce to a single output.

By using pull-up and pull-down resistors as before we can create eight codes with three switches.

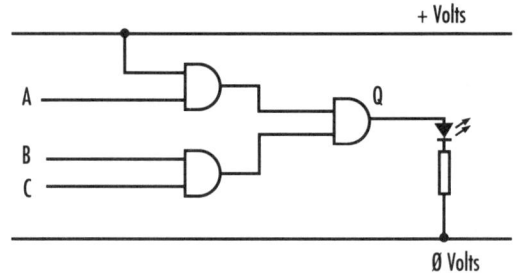

0	0	0
1	0	0
0	1	0
0	0	1
1	1	0
0	1	1
1	0	1
1	1	1

8 Possible codes

Logic Codes

An alternative way is to use all resistors as pull-downs so that all switches in the on position are high. To get different codes we use different types of gate.

Here is where a good knowledge of logic statements is very useful.

For example, to use a code of 1,1,0,0 we would need an AND gate and a NOR gate (output high only when both inputs are low) and a second AND gate to reduce to a single output.

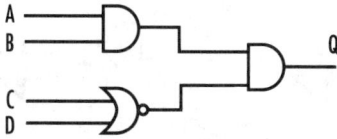

Code 1,1,0,0

To use a code of 1,0,1,0 or 1,0,0,1 or 0,0,1,1 we use the same gates because we still have two at high and two at low.

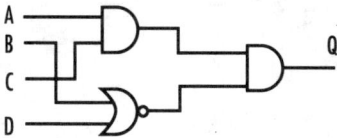

Code 1,0,1,0

The problem arises when we have a three and one situation: 1,0,0,0 or 0,1,1,1 etc.

Here we can use our friend the NOT gate or inverter to help with the odd-man-out.

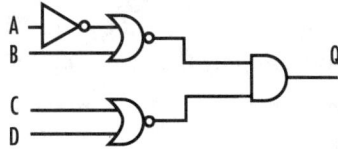

Code 1,0,1,0

Can you work out how many possible codes there could be for four switches or even five or six?

Logic Gate Experiments

The Logic Game

This board game requires a knowledge of logic gates. If you get stuck you can look at the mini-dictionary. The more you play, the more familiar the gates will be and the quicker you will get. You can play the game on your own or with friends.

As you can see, the board consists of a course of gates that you have to go through. To get through a gate you have to throw the correct inputs according to what output is needed.

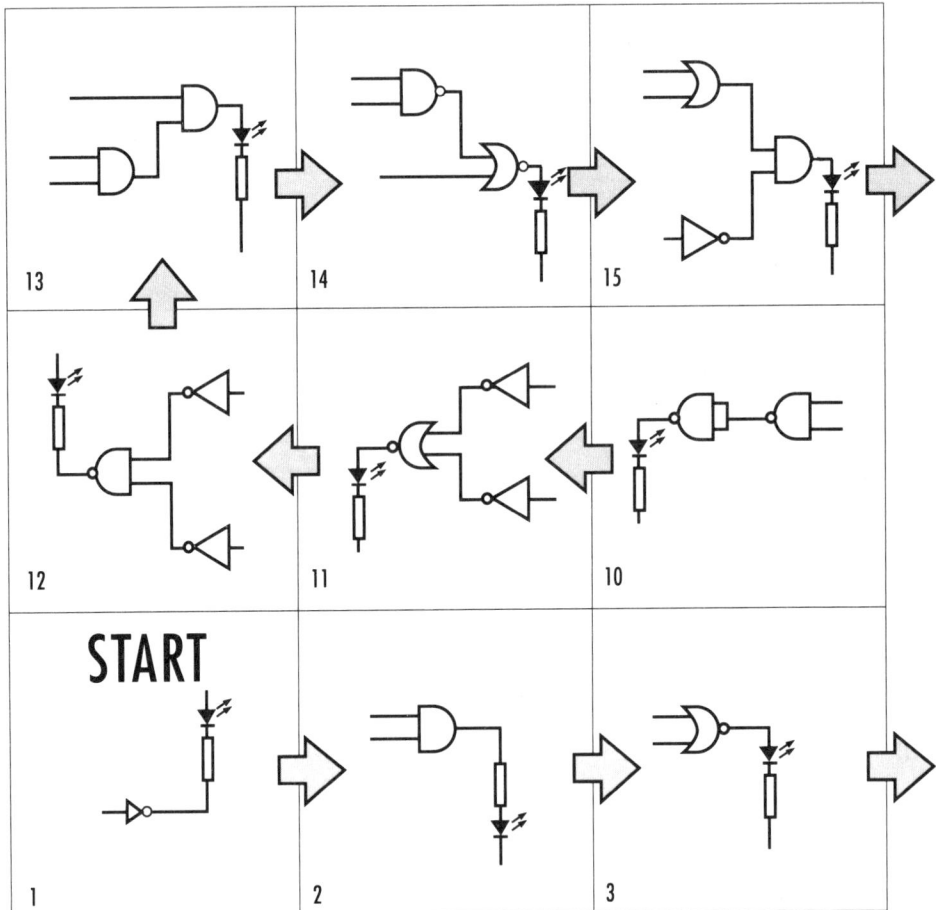

If the LED and resistor are going up towards + volts then the output needs to be low (sinking). If they go down towards 0 volts, the output must be made high (sourcing).

You need three dice (even numbers high, odd numbers low) or you could use coins (heads high, tails low). Use as many dice or coins as the gate has inputs. If you get it right, move on. If you get it wrong, it's the next player's turn.

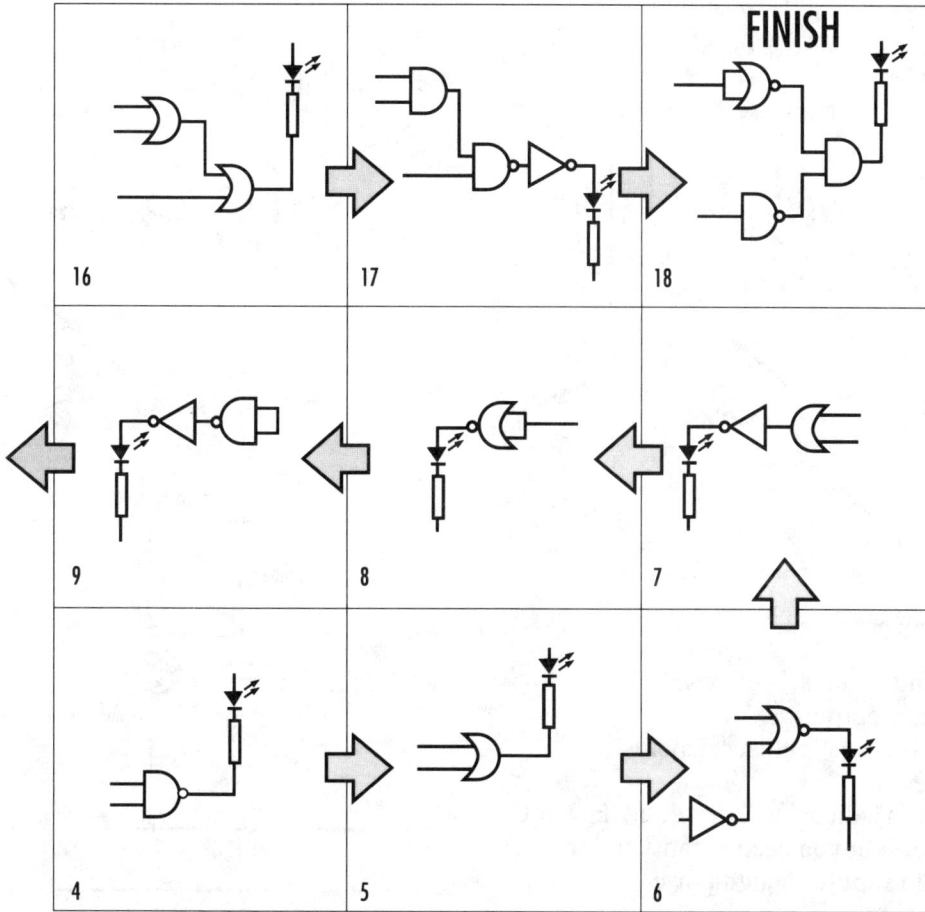

Making a Security Code

Design and make an electronic device which requires a digital code. It could be a security lock for a door or box. It could be a type of Mastermind game where codes can be set and altered for a friend to try to crack.

Points to think about:

1 What type of switch will you use?
   - Slide, rocker, toggle, push-button, or possibly a membrane panel? (*See* **Electronics**.)

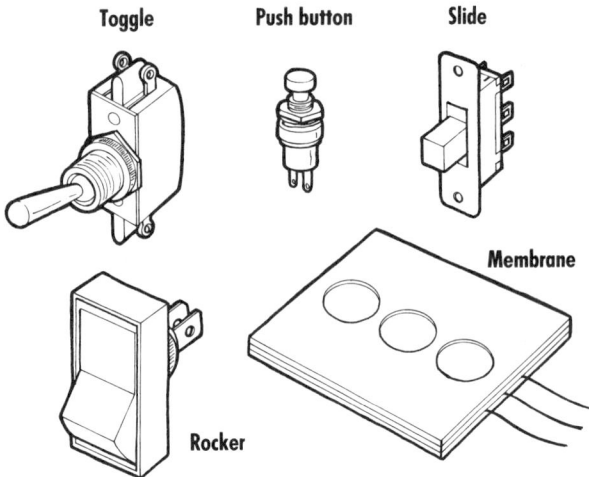

Toggle   **Push button**   **Slide**

Membrane

Rocker

2 How can the code be changed easily if necessary?
   - Plugs and sockets, switches?
3 What type of output?
   - LEDs, buzzers, bulbs, solenoids, motors?

As you know, a logic IC will drive an LED but for a bigger output you need a transistor or even a relay to supply enough power.

A relay is an electromagnetic switch which is turned on by a low power circuit. It is used as an *interface* between a low-power and a high-power circuit.

The output from a transistor driven from an IC will energize the coil of the relay, turning on a switch which will operate a larger device such as a motor or solenoid.

A diode is needed here to protect the transistor.

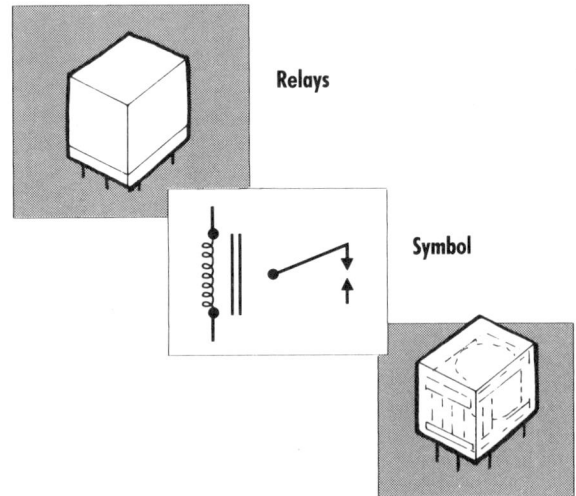

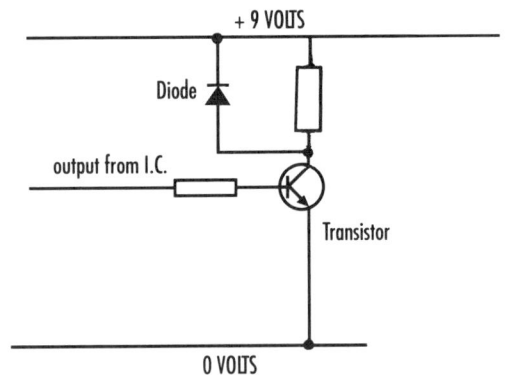

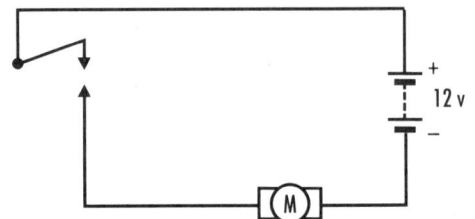

Relays

Symbol

+ 9 VOLTS

Diode

output from I.C.

Transistor

0 VOLTS

+
12 v
−

M

7 Multivibrators

Monostable

A **monostable** is a device which has only one stable state. This means that whatever we do to the circuit it will always revert to the same state.

We use this to our advantage by creating time-delays to turn something on or off after a timed period.

To make a monostable we need two extra components added to a pair of inverters: a **variable resistor** (VR) and a capacitor.

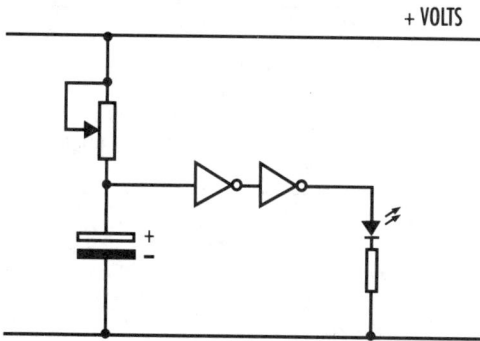

The variable resistor, sometimes called the **potentiometer** or 'pot', does just what its name suggests. The resistance can be varied by adjusting a knob in the case of large pots or using a small screwdriver in the case of 'preset' pots.

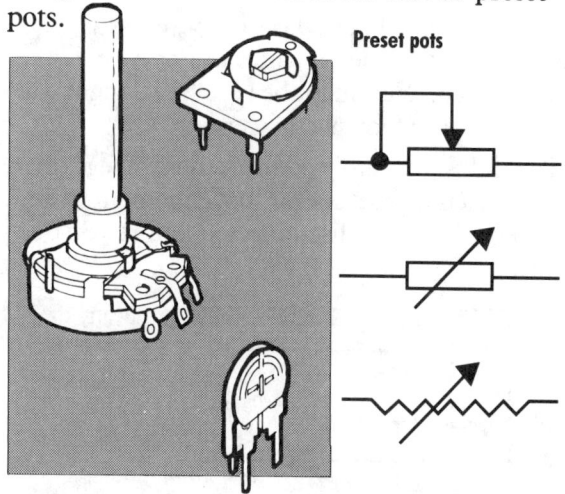

Preset pots

The capacitor is a component that acts much like a tiny rechargeable battery. It can store electricity. The larger its value, the more it can hold and its actual size usually increases with value. (*See* **Electronics**.)

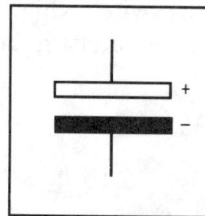

ELECTROLYTIC CAPACITOR SYMBOL

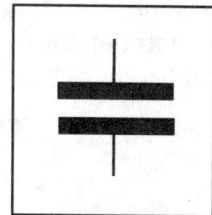

NON-ELECTROLYTIC SYMBOL

Its value is measured in microfarads (μF). Most capacitors with a value of $1\mu f$ or more are electrolytic and must be connected the correct way round in a circuit.

Most capacitors of less than $1\mu F$ are non-electrolytic and can be connected either way round.

Multivibrators

From the circuit diagram we can see how current can pass through the VR and into the capacitor. When the capacitor is full, the current will overflow into the gate input, making it high.

To start a timed period properly, we must make sure that the capacitor is empty or discharged. We cannot be sure how long it will take if it is already partly charged.

To do this we simply connect the two legs of the capacitor together for just a moment. The battery should be disconnected before you do this.

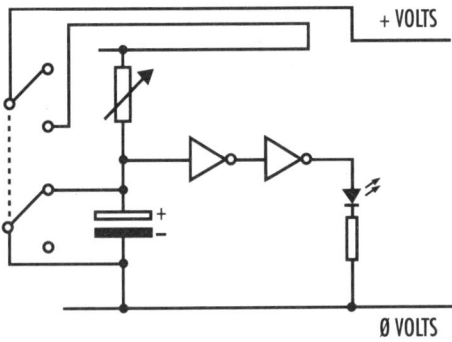

To kill two birds with one stone you could use a double-pole, double-throw switch which would 'short out' the capacitor and turn the circuit off at the same time.

Experiment 9

Using a prototype board or your own logic board, set up a monostable circuit and try a variety of capacitors and resistor values. If you have an ohmmeter you can read the resistance values of the VR. If you haven't, use fixed resistors of known value. Time each set-up and record your results either as tables, graphs or both.

Resistor Value	Capacitor Value	Time

or

Astable Multivibrator

An **astable** is a circuit which has no stable state. It keeps changing from high to low and back again.

If an LED was used as an output from an astable it would flash on and off continuously.

Again, we use a pair of inverters plus a VR and a non-electrolytic capacitor.

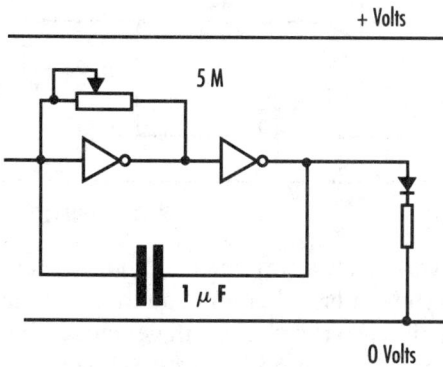

Experiment 10

Set up the astable multivibrator on a prototype board or your own logic board.

Use the component values which are given in the circuit diagram.

Adjust the VR and see what happens.

As the resistance is reduced, the flashing rate is increased. The flashing rate will continue to increase as you adjust the VR until the LED appears to stop. In actual fact it hasn't stopped: it is going too fast for our eyes to see the ons and offs.

One complete on and off is called a *cycle*. The number of cycles in a second is called the *frequency*. This frequency or rate is measured in *hertz* (Hz). If the LED is flashing on and off 10 times per second, this is given as 10 Hz.

If you now slow the frequency down so that each cycle is about five seconds (0.2 Hz) you can time the exact period that the LED is on and how long it is off.

You should find that the on time is exactly the same as the off time.

Multivibrators

The LED also switches on and off very cleanly. It doesn't creep on slowly and then fade off. In other words, the circuit changes instantly from high to low and back again. If we show this as a graph of logic levels against time, we have what is called a *square wave*.

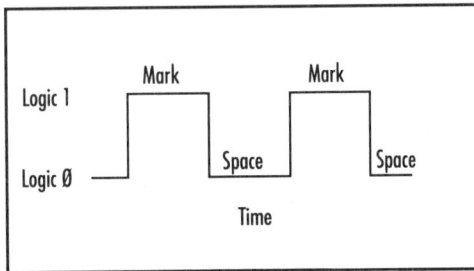

The time that the output is high is called the *mark*. The time that it is low is called the *space*. When we compare the on time with the off time we call this the mark/space ratio. In this case they are the same, so the mark/space ratio is 1:1.

If the LED was on for less time than it was off, the wave would look like this.

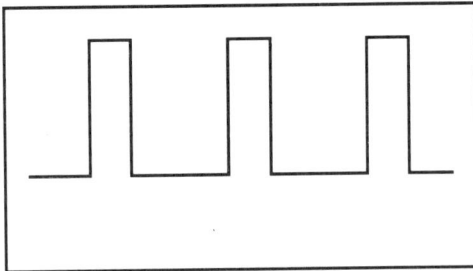

What would the wave look like if the LED was on for longer than it was off?

Experiment 11

Set up another astable but use a loudspeaker as the output.

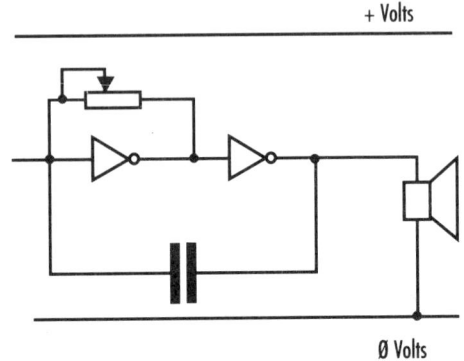

On every cycle a soft click should be heard from the loudspeaker. As you increase the frequency, so the clicks will get closer together and may even sound like a deep buzz.

Now change the component values to much smaller ones and notice the difference in the tone.

What we have now is a one-note organ. To have more notes we simply add more variable resistors as shown and tune each one to the required note. Preset VRs are most suitable for this (and cheaper).

To make the sound louder, we add a resistor and transistor to the output as before.

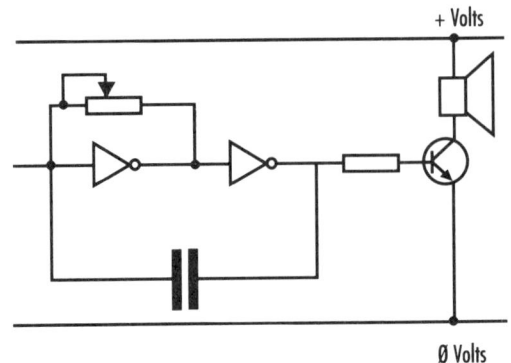

Making an Organ

Design and make a one-octave organ.

Points to think about:

1 How many notes do we need?
2 How do we change the notes to play a tune?
   - Push-button switches, keyboard switches, a probe, membrane panel switches?
3 Circuit construction:
   - PCB, matrix, strip board?
4 Design for a case:
   - Make a suitable box from plastic or wood?
   - Electrical trunking?
   - Rainwater pipe? This could be round or square.
   - Use and adapt a ready-made box?

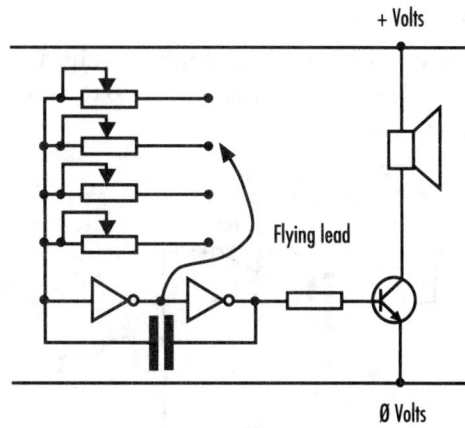

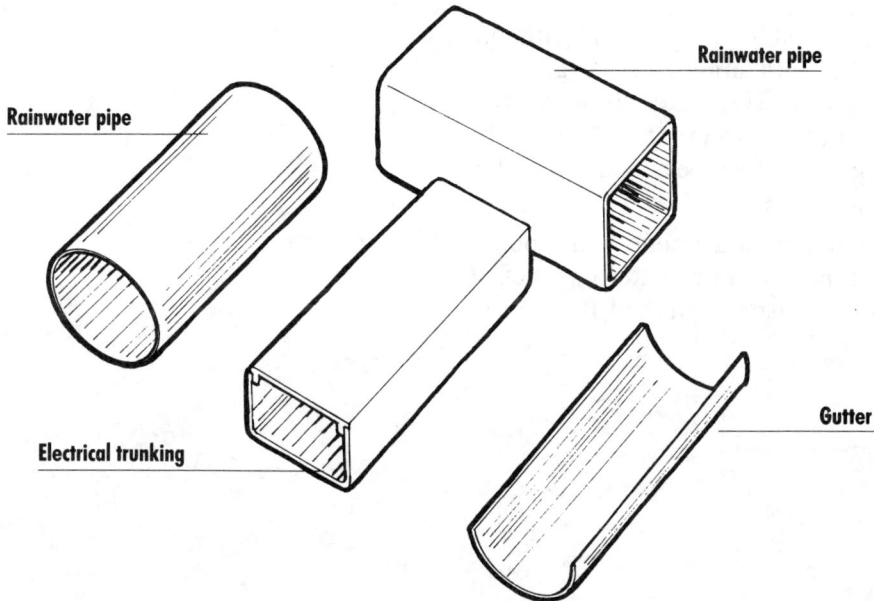

Multivibrators

Combining Astables

Experiment 12

Set up an astable with component values as for a single tone but take the output to the input of an AND gate.

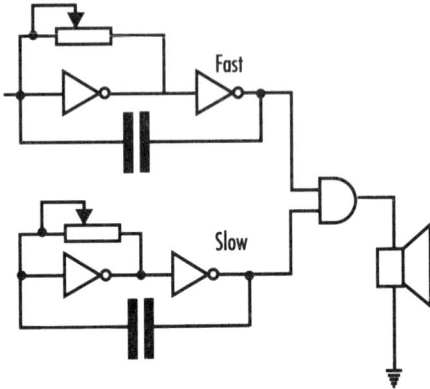

Set up a second astable (use the spare gates on the same IC) but with larger values to give a frequency of about 1 Hz. Take this output to the other input of the AND gate. The output of the AND gate can then be connected to the loudspeaker as before.

With possible minor adjustments you should get an on/off tone. The pitch and frequency of the tone can be adjusted with the VRs.

Experiment 13

Let's go one stage further and set up three astables: astable 1 with a low tone, astable 2 with a high tone and astable 3 at about 1 Hz to act as an automatic switch between the two tones. Your result should be much like that of a police car siren.

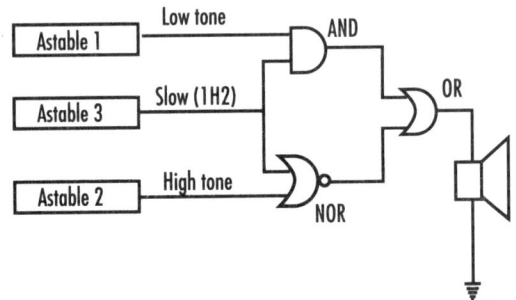

Bistables

A **bistable** or **flip-flop** is a circuit that is stable in either state. It has two inputs and two outputs.

We can make a bistable using two NAND or two NOR gates.

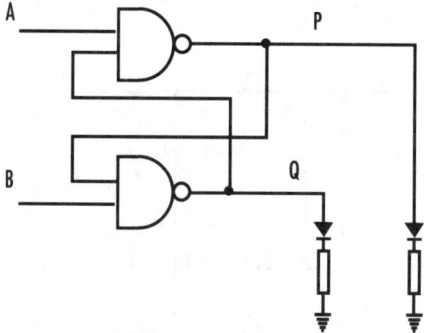

Experiment 14

Set up a bistable with LED outputs using a pair of NAND gates.

Before we start, let's remind ourselves that for a NAND gate the output is low only when *both* inputs are high. Or, to put it another way, the output is always high if there is at least one input low.

Keeping this in mind, and using the truth table, try to predict the outputs before you actually change the inputs on each line and see if you are correct.

Start by making A low and B high.
Then make A high as well.
Next make B low.
Then make B high again.
Next make both low.
Then make both high.

Carry on changing inputs and recording outputs. Sometimes change one input at a time, sometimes change them both.

What you should notice is that the output stays the same even after one of the inputs has changed. This system is said to have *memory*. Also, in some cases you could not predict the outputs accurately. When?

With a little imagination this circuit could be adapted to make a quiz-game indicator where the first player to hit the button makes a light come on. If the second player then hits his or her button it has no effect. This is the basis of the systems seen in TV games.

A	B	P	Q
0	1		
1	1		
1	0		

Truth table

8 A 555 Timer

Monostable

The 555 is a very cheap and popular IC. It has so many uses that many books have been written about it alone. Thousands of interesting projects rely on it in order to function.

The 555 comes in an 8-pin DIL package. The diagram may look much more complicated than previous ICs, but just remember the 'black box' theory, obey the basic rules and you can't go wrong.

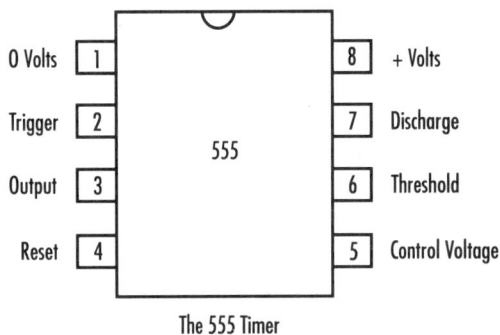

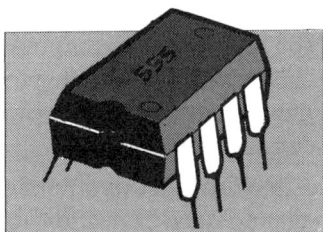

0 Volts	1			8	+ Volts
Trigger	2		555	7	Discharge
Output	3			6	Threshold
Reset	4			5	Control Voltage

The 555 Timer

It can be used in two basic ways: as a monostable and as an astable multivibrator.

In its monostable mode we can use it as a timer. We can set a device to come on after a period of time or we can set it to turn a device off after a period of time, as with the time-delay circuit in the previous chapter. The timed period can be varied very easily as before.

The circuit diagram shown will switch a device off after the timed period. To make it switch on, simply place the LED in the dotted position. As an astable it acts as a clock, giving out regular pulses (on, off, on, off) in a square wave. The frequency can also be varied very easily.

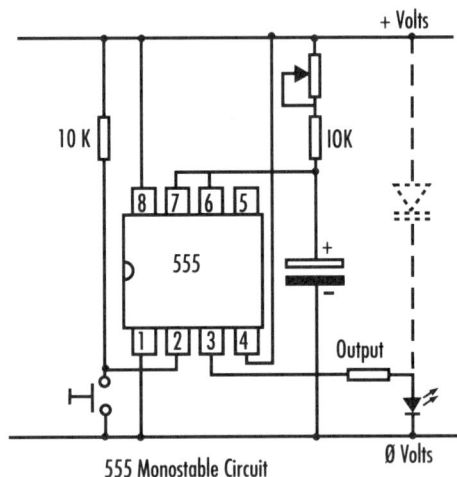

555 Monostable Circuit

Again, for both types of circuit we need other components: the VR and the capacitor.

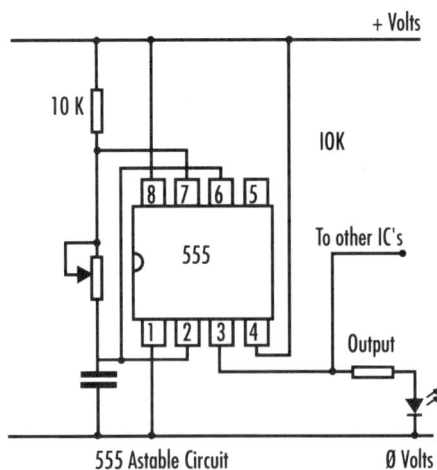

555 Astable Circuit

Making a Clock

555 Astable Multivibrator

This project can be used as a clock, giving out a square wave suitable to drive other ICs in later projects.

Points to think about:
1 In order to drive other ICs it must share their power supply, so suitable connections are needed.
2 Its output needs to be connected to the input on other circuits, so again a suitable connection method is needed.
3 The frequency needs to be easily adjusted and **calibrated** to a reasonable degree of accuracy. To calibrate such a device, the control knob or handle needs a pointer or mark of some kind on it. This is used to point to numbers or dots which will show the frequency in hertz. Calibration is done by trial and error: by adjusting the control, counting the frequency and marking the surface of the case or package accordingly. Once calibration is done it should be possible to set the clock to any frequency within its range quite easily.

The problem you will encounter quite quickly is that, at a certain frequency, it will be too fast for you to count.

Get as far as you can and we will come back to difficult calibration in the next chapter. As usual, a suitable package needs to be thought about considering time, cost, size, shape, etc.

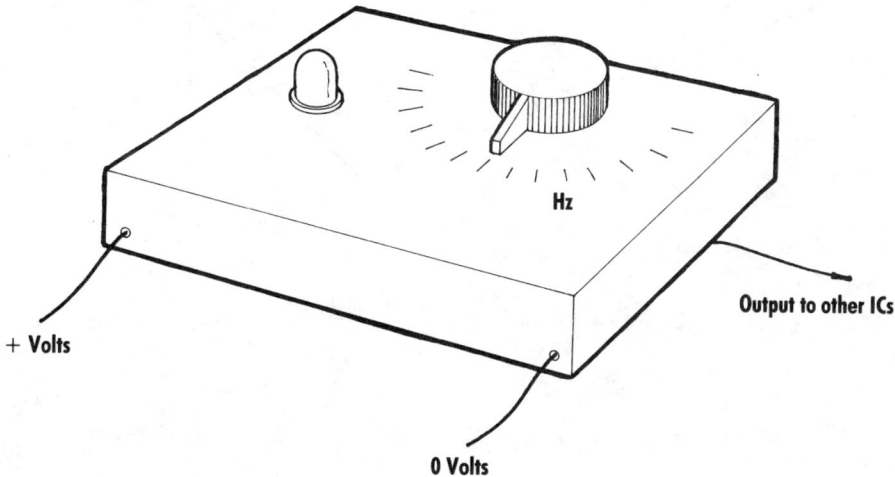

Hz

Output to other ICs

+ Volts

0 Volts

Making a Timer

Monostable Timer

Using the 555 monostable circuit, design and make a timer that could be used for a situation of your choice.

Points to think about:

1 Decide exactly the function of the timer. Is it to come on after a timed period or to turn off?

2 Decide on an output:
  – LED, bulb, buzzer or perhaps another circuit?
3 Does it need to be easily adjusted for various time delays? If so, then careful calibration is needed as before.
4 Again, think about packaging.

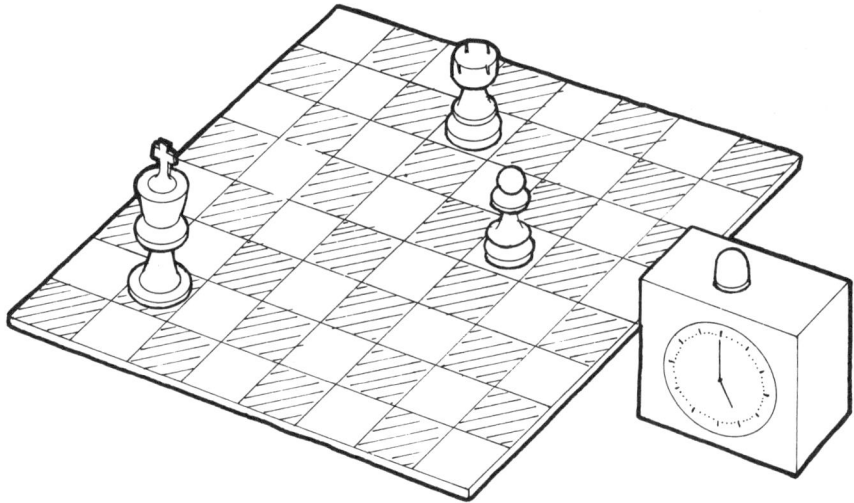

Questions

Test Yourself So Far

1 Explain the following terms:
   Mark
   Space
   Frequency
   Hertz
2 Which LED would be lit in this circuit? Explain.

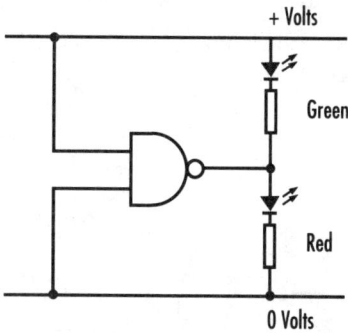

+ Volts

Green

Red

0 Volts

3 Give two ways of adjusting the frequency of an astable.
4 There are five important mistakes in this circuit. Can you find them?

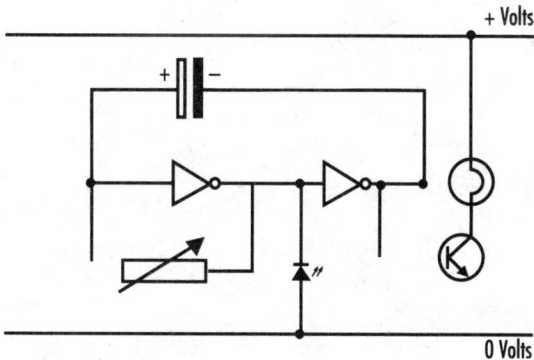

+ Volts

0 Volts

5 What is a relay?
6 Explain what is meant by the following:
   Monostable
   Bistable
   Astable

7 Can you predict the outputs of this bistable? Explain.

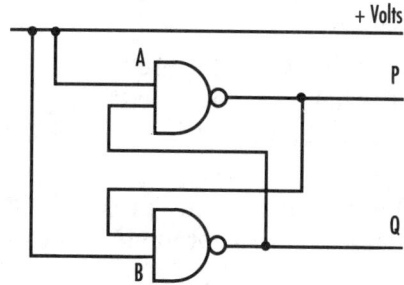

+ Volts

A

P

B

Q

8 Explain what is meant by pull-down and pull-up resistors.
9 How do we stop gate inputs from floating?
10 What is meant by calibration?
11 Work out the codes to give a high output in the following:

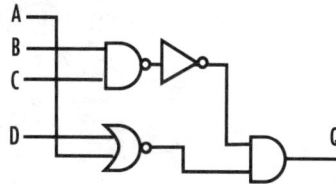

A
B
C
D
Q

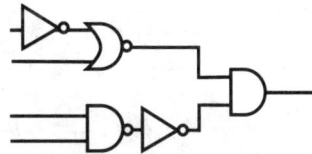

9 Dedicated ICs

So far we have been using basic integrated circuits, adding other components and combining different ICs to perform various functions.

In this chapter we deal entirely with **dedicated ICs**. They have been designed to do a particular job and they will work perfectly well if connected correctly.

Each IC can be made into a project on its own or you may want to incorporate it into other circuits.

Each project can be made using a PCB, matrix board, etc., but for a permanent job the prototype board or your own logic board would be unsuitable.

You are well advised to *model* your circuit on such a board before you produce the real one. Remember the hints on page 14.

If you have any trouble, follow the fault-finding guide on the next page.

This is take-off time. The sky's the limit. Good luck.

Fault-finding

Probably the most frustrating part of any electronic project is when it doesn't work straight away.

A lot of time can be saved and disasters avoided by making certain checks and double-checks during design and construction. Some hints are given on page 14.

There are lots of things to check when fault-finding, but what to check *first* really depends on the symptoms. There is no substitute for experience but the following guidelines may be helpful; look also at the fault-finding guide in the **Electronics** book.

Remember: faulty components are very rare. If your circuit appears to be dead, you must first check out the power supply. If you use a 2-LED logic probe, connect the supply leads to the battery and first touch the probe to the + volts rail on the circuit, then the 0 volts rail. If nothing happens with either or both, the fault must be between the supply rails and the battery. Is the battery flat? The battery snaps could be faulty.

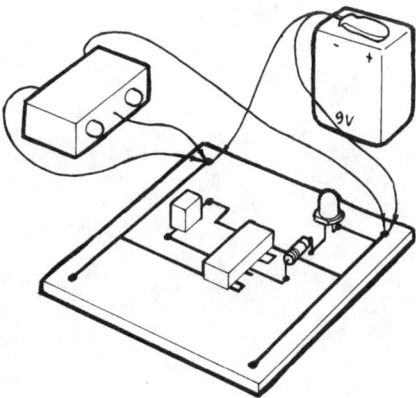

If you have got power to the circuit but it still doesn't work, turn off and check that all components are in the right place and the correct way round. Make a close visual check of each soldered joint.

Next check that each leg of each component is actually making contact with the track. Use a continuity tester and probe between the actual leg and the track next to the joint. Check for continuity along the lengths of the tracks and wires.

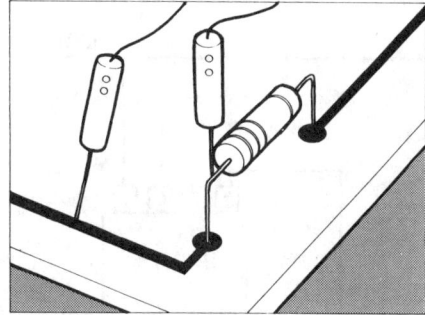

Finally, turn on again and check with the logic probe that each point on the circuit that should be high *is* high and that each definite low point *is* low.

If you still have no luck then ask your teacher to make a check for you.

Dedicated ICs

LED Flasher

LM3909N (An 8-pin DIL IC)

You would be right in thinking that you already know how to make an LED flash but this IC is rather different.

1 It only uses a single 1.5 volt C-size cell.
2 The cell should last for several months even with constant use.
3 It produces an unusual square wave, where the space is much larger than the mark. This gives a flash followed by a longer delay, much like the flashing yellow beacons around roadworks.

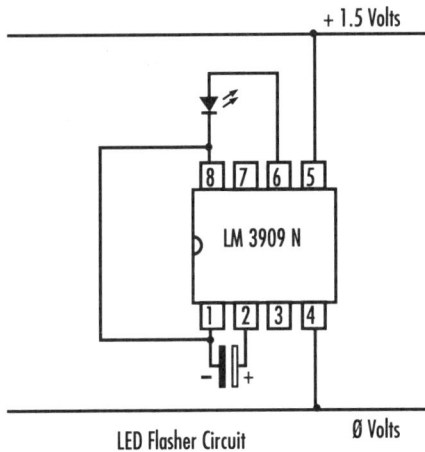

Wave from LED Flasher Circuit

Possible projects could be a dummy alarm for a car. Potential burglars might well stay clear if the LED was mounted in a 'professional' box and fixed in full view on the dashboard. The same could be used for a shed or even a house.

A keyhole indicator? Fixed in a convenient place it would help find the keyhole if the door wasn't well lit.

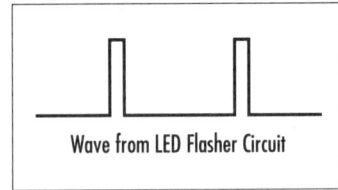

LED Flasher Circuit

Melody Maker

UM86 Melody Generator

This is a simple IC requiring one or two 1.5 volt cells to power it. It comes in a 3-leg package much like a small transistor. Its output is a small piezo transducer, QY13P.

Various tunes are available.

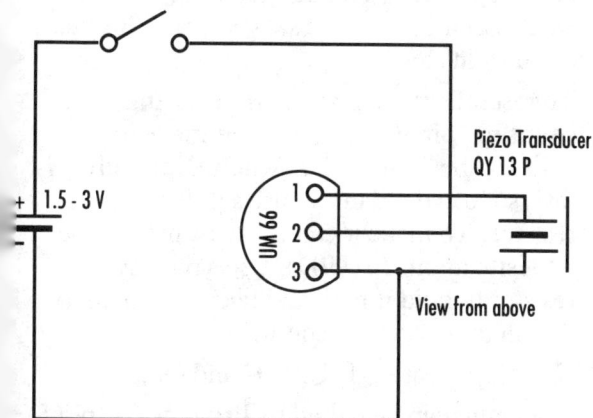

If you want a louder output, a simple amplifier can be added as shown. To make it work you simply connect the battery. It will play the tune once only and then stop. The battery must be disconnected and reconnected if you want to repeat it. This could, of course, be done with a simple switch.

Suitable projects could be simple greetings or novelty cards; a replacement for a doorbell; automatic 'welcome' tune every time someone comes through a door; musical money-box, etc.

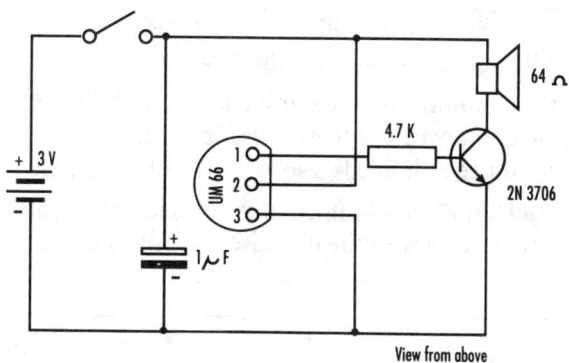

Piezo Transducer
QY 13 P

1.5 - 3 V

UM 66

1
2
3

View from above

3 V

UM 66

1
2
3

4.7 K

64 Ω

2N 3706

1 μF

View from above

Dedicated ICs

Binary Counter

4516 Binary Up/Down Counter

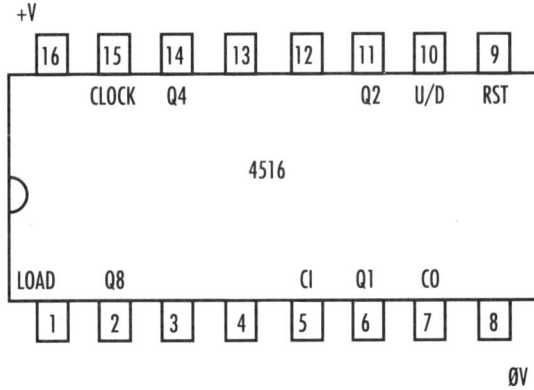

+V

16	15	14	13	12	11	10	9

| | CLOCK | Q4 | | | | Q2 | U/D | RST |

4516

| LOAD | Q8 | | | CI | Q1 | CO | |

| 1 | 2 | 3 | 4 | 5 | 6 | 7 | 8 |

ØV

This IC will count up to 15 in binary.

The output is a 4-LED display.

It can count forwards or backwards. This is an ideal opportunity to put your astable project to good use as it needs a square wave to drive it.

Such an IC would normally be used as part of a much more complicated circuit, but as a project you could make it an aid for learning the binary counting system.

If you want to count higher than 15, a second 4516 IC can be added. We call this *cascading*. The carry-out pin 7 (CO) is connected to the carry-in pin 5 (CI) of the next IC. This would count up to 128. You can add even more if you want. Each IC uses the same clock. This keeps them all in step.

To reset the system at any time during counting, pin 9 (RST) must be made high. This can be done with a switch. Normally it is low, so to avoid damage use a pull-down resistor. To make it count backwards do the opposite to pin 10 (U/D). Use a pull-up resistor to hold it high and take it low with a switch to reverse the count.

The outputs are Q1, Q2, Q4 and Q8.

The count can be halted by bringing CI (pin 5) high. (Don't forget the pull-down.)

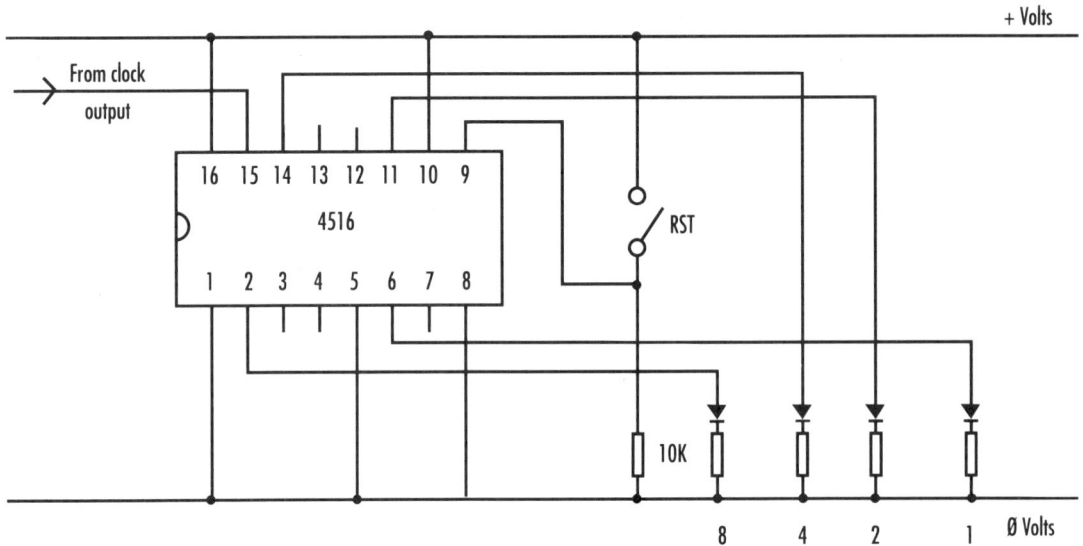

Binary Counter

Fact File

In IC manufacture the slightest amount of contamination could ruin thousands of ICs and cost the company thousands of pounds. An IC laboratory is about a hundred times cleaner than a hospital operating theatre and the technicians dress like surgeons.

Divide by 10 Counter

4017 Divide by 10

This IC has 10 outputs (0 to 9). Using a clock input, each output will come on and then go off in that order. If you connect an LED to each output, the result would be a 'chain' or 'ripple' effect which can be made to go faster or slower.

It also has a divide by 10 output (pin 12) which gives out a pulse on the tenth count. The reset method is the same as the previous circuit (4516).

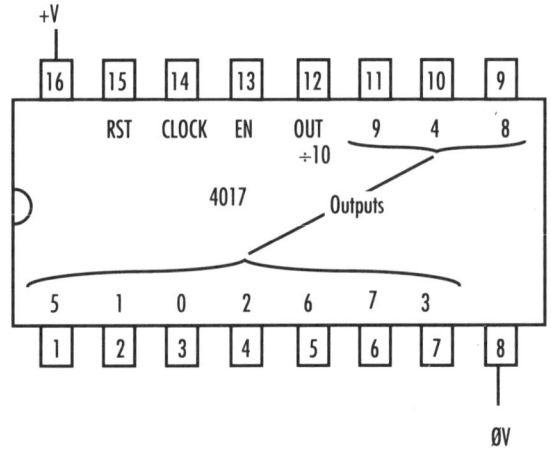

Alternatively you can have an automatic reset on any number you wish. If you only want to count up to, say, 4, you simply connect the 5th output (output no. 4) to the reset pin.

As with logic ICs, if you want a larger output such as a buzzer or motor you will need the usual resistor/transistor set-up.

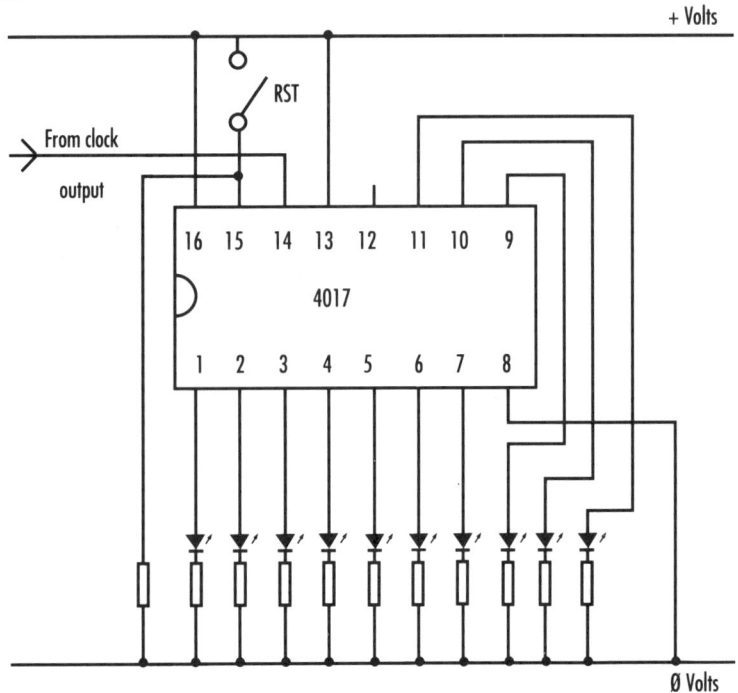

Divide by 10 Counter

Possible uses could be LED dice. (Count to six, reset on seven.) Set the clock fast enough so that all the LEDs appear to be on. Use a switch to make the enable pin (13) high (don't forget the pull-down resistor), which will halt the action at whatever LED is lit. This is then a 'random' choice.

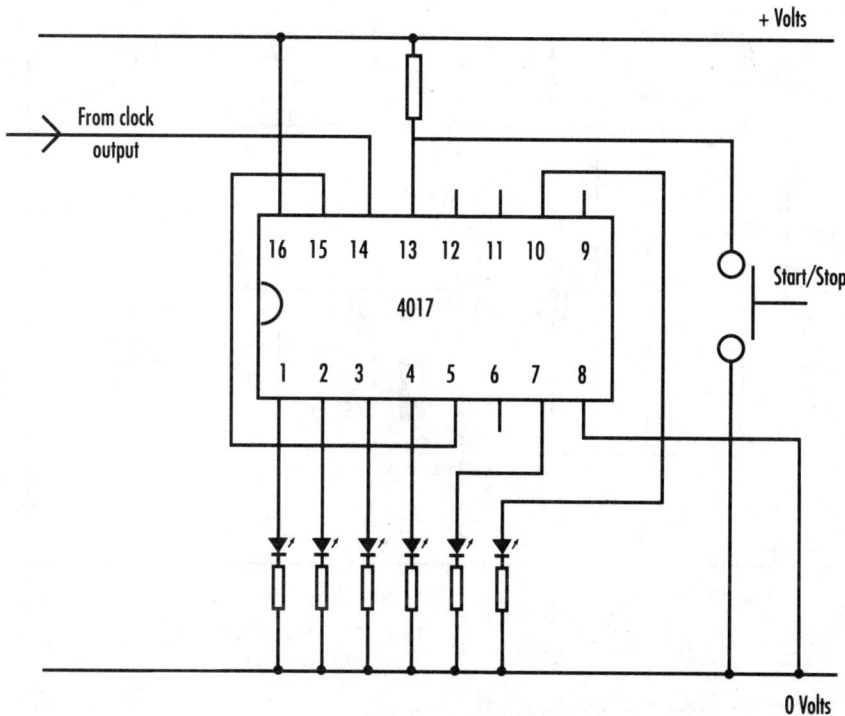

By combining this with tuned astables you could compose your own tune or jingle.

This IC will help with your calibration of the clock circuit. When it's too fast to see, connect it to the 4017, connect a LED to pin 12 and multiply your count by ten.

The possibilities for this IC are almost endless if you want an automatic sequence of events.

Dedicated ICs

Personal Radio

Tuned Radio ZN416

Here you can make your own personal radio.
The circuit is very simple, as you can see. It
runs on a single 1.5 volt cell.

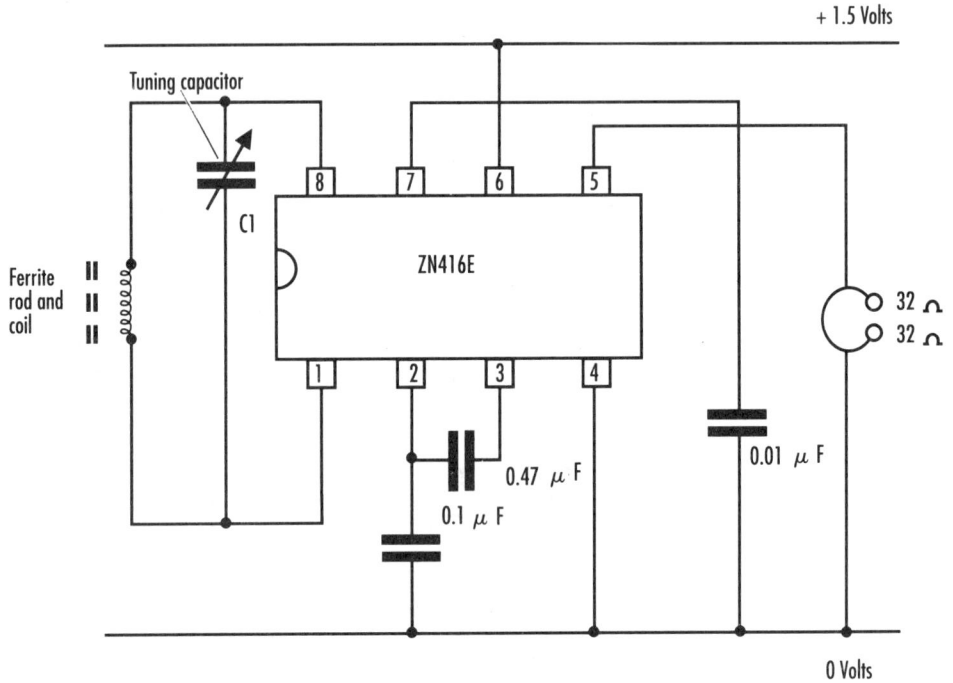

+ 1.5 Volts

Tuning capacitor

C1

8   7   6   5

ZN416E

Ferrite
rod and
coil

32 Ω
32 Ω

1   2   3   4

0.47 μ F

0.1 μ F

0.01 μ F

0 Volts

Two components that you may not recognize
are needed: a variable capacitor for tuning and
a **ferrite rod** to receive the radio signals.

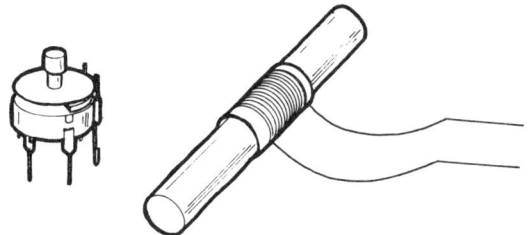

Personal Radio

The earphones must be connected in series as shown. They will almost certainly be wired for stereo when you buy them and so you must convert them. Either buy a stereo/mono adaptor or cut the plug off and reconnect the wires as follows:

1 Strip about 2 cm of the outside sleeve from both wires.
2 Peel back the copper sleeves.
3 Strip about 1 cm from each of the inner wires.
4 Twist the copper sleeve from one to the inner wire of the other and tape the joint.

This then leaves you with just two connections to go to the PCB.

For best reception, and to avoid interference, you should observe the following points when constructing your circuit:

1 The 'earthy' side (fixed side) of the tuning capacitor C1 should be connected to pin 8.
2 The tuning capacitor and its leads or tracks should be as far away as possible from the battery, headphones and their leads.

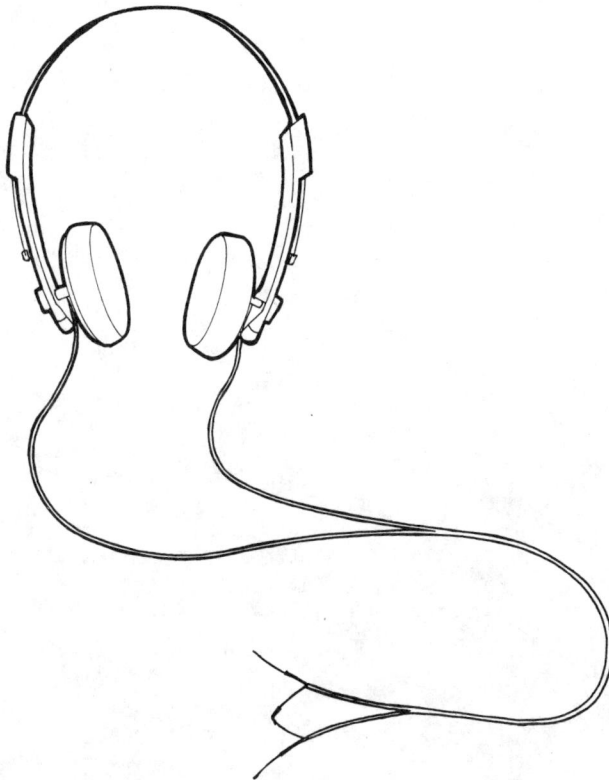

AND gate The output is high only when both inputs are high.

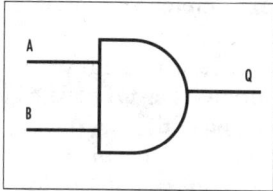

A	B	Q
0	0	0
0	1	0
1	0	0
1	1	1

Astable A circuit which is stable in neither state. Its output is a square wave.

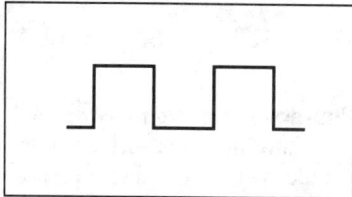

Bistable A circuit which is stable in either state. It can be made from NAND or NOR gates.

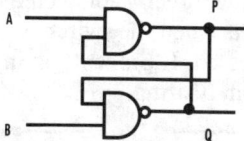

Bit A single piece of information in the form of an off or on, high or low, positive or negative, 1 or 0.

Black box A device which we know what it is capable of and how to use it but we don't understand how it works. Nor do we need to.

Byte Series of eight bits. Kilobyte: one thousand bytes. Megabyte: one million bytes.

Calibration Marks, dots, lines, numbers to represent a value. Some kind of indicator usually points to them.

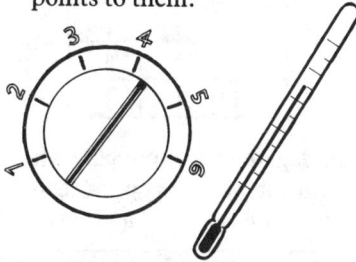

CMOS A family of ICs standing for Complementary Metal Oxide Semiconductor.

Dedicated ICs Integrated circuits designed to do a particular job, e.g. amplifier, radio, etc.

Digital Working on a system of on and off, high and low, yes and no, 1 and 0, north and south, + volts and 0 volts. No values in between the extremes.

DIL Stands for Dual In Line and refers to the type of package that contains an integrated circuit. It has two (dual) rows of pins or legs which are parallel (in line).

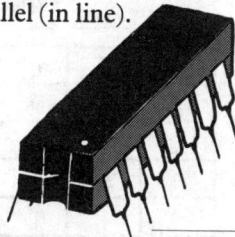

Earthing Touching a metal conductor which is connected to earth in order to get rid of static electricity in our body before handling a CMOS IC.

Feedback resistor Carries a small amount of current from an output back to the input of a gate or system of gates. Used as a latch in the electronic switch circuit.

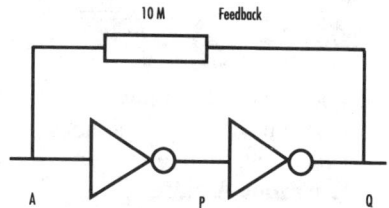

Ferrite rod Black rod with a coil of wire around it used as an aerial for a radio.

555 Timer A very popular multi-purpose IC which is mainly used as a monostable or an astable. It comes in an 8-pin DIL package.

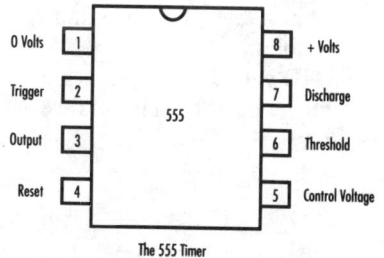

The 555 Timer

Floating Describes an input to a gate which is not connected to either + volts or 0 volts. It can flick from high to low very quickly and uses up a lot of current. This must be avoided with CMOS ICs. Unused inputs should be connected to the nearest convenient supply rail.

Mini-Dictionary

IC Stands for integrated circuit. A complex circuit containing thousands of components reduced to microscopic size on a 'chip' of silicon.

Inverter *See* NOT gate.

Logic gate A circuit which has inputs and an output. Each different gate does a different job and has its own symbol.

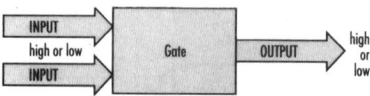

Logic level High or low, + volts or 0 volts, 1 or 0; sometimes called a logic state.

Logic probe A piece of test equipment used to indicate the logic level at particular points in a circuit.

Monostable A circuit which is stable only in one state. Used where a time delay is needed.

Multivibrator General name for a two-stage switching circuit. There are three kinds: monostables, bistables and astables.

NAND gate The output is high when one or both inputs are low.

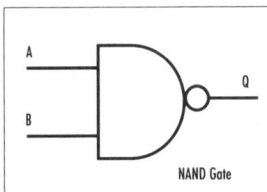

A	B	Q
Ø	Ø	1
Ø	1	1
1	Ø	1
1	1	Ø

NOR gate The output is high only when both inputs are low.

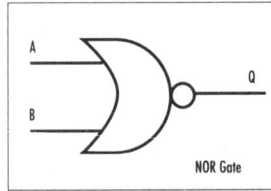

A	B	Q
Ø	Ø	1
Ø	1	Ø
1	Ø	Ø
1	1	Ø

NOT gate One-input gate whose output is always opposite to the input. Also called an inverter.

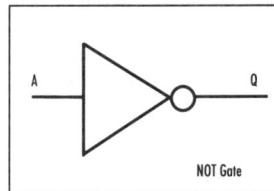

A	Q
Ø	1
1	Ø

OR gate The output is high when one or both inputs are high.

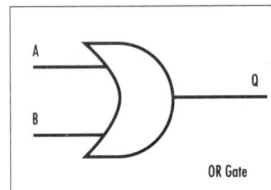

A	B	Q
Ø	Ø	Ø
Ø	1	1
1	Ø	1
1	1	1

Package The type of case or container that a component is set in.

Potentiometer *See* Variable resistor.

Preset *See* Variable resistor.

Prototype board A plastic board consisting of parallel rows of tiny sockets into which various components can be inserted in order to model a circuit or part of a circuit.

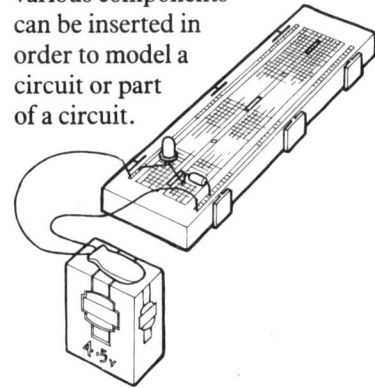

Pull-down/pull-up resistors A resistor in series with a single pole switch. The switch is an easy way of changing from high to low. The junction of the two is usually connected to a pin on an IC. It prevents + volts from being directly connected to 0 volts when the switch is pressed. It prevents an input from floating.

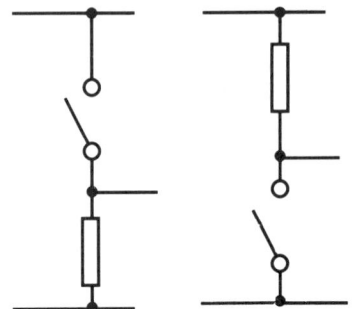

Sinking When the output of a gate is at logic level 0. Current flows in.

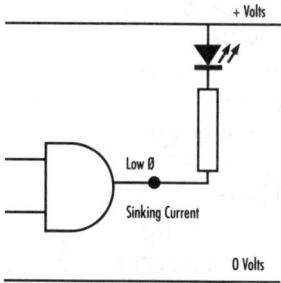

Sourcing When the output of a gate is at logic level 1. Current flows out.

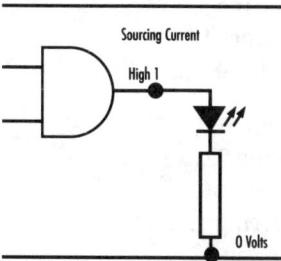

Static *See* Earthing.

Transducer General name for a component which changes one form of energy into another. The transducer used in the melody-maker circuit changes electrical energy from the IC to sound energy: music.

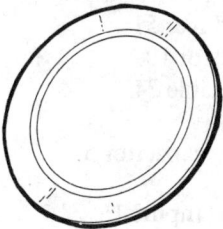

TTL A family of integrated circuits. It stands for Transistor Transistor Logic.

Tuning capacitor A variable capacitor which can be adjusted in a similar way to variable resistors.

Variable resistor Sometimes called a potentiometer. Resistance can be varied using a control knob, or a small screwdriver in the case of a preset.

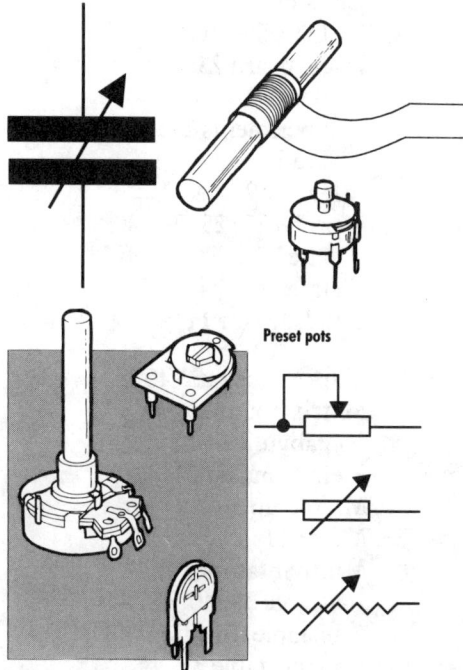

Preset pots

Index